The Andrew R. Cecil Lectures on Moral Values in a Free Society

established by

The University of Texas at Dallas

Volume VIII

TRADITIONAL MORAL VALUES
IN THE AGE OF TECHNOLOGY

Traditional Moral Values in the Age of Technology

HANS MARK
TOM L. BEAUCHAMP
JESSE P. LUTON, JR.
MARTIN E. MARTY
ANDREW R. CECIL

With an Introduction by 68924
ANDREW R. CECIL

Edited by
W. LAWSON TAITTE

The University of Texas at Dallas
1987

Library of Congress Catalog Card Number 87-050289
International Standard Book Number 0-292-78098-2

Distributed by the University of Texas Press,
Box 7819, Austin, Texas 78712

FOREWORD

The Andrew R. Cecil Lectures on Moral Values in a Free Society have become an important tradition at The University of Texas at Dallas since the series was established in 1979. Each year the University invites to its campus leading statesmen and scholars of national prominence so that they may share their ideas with the academic community and the general public. The continuing theme of the series is the system of moral values on which our country was built, and the Lectures offer a unique forum for the analysis and debate of these values. In offering the Lectures on Moral Values in a Free Society, the University is fulfilling an important obligation and trust.

This program is named for Dr. Andrew R. Cecil, the University's Distinguished Scholar in Residence. When Dr. Cecil served as President of The Southwestern Legal Foundation, his innovative leadership brought that institution into the forefront of continuing legal education in the United States. Upon his retirement from the Foundation as its Chancellor Emeritus, Dr. Cecil was invited by The University of Texas at Dallas to serve as its Distinguished Scholar in Residence, and the Cecil Lectures were instituted. They appropriately honor a man who, throughout his career, has been concerned with the moral foundations of our society and has stressed a belief in the dignity and worth of the individual.

The eighth annual series of the Cecil Lectures was held on the campus of the University on November 11-13, 1986. The theme of this eighth series, "Traditional Moral Values in the Age of Technology," has a special relevance to the University at this point in its history. Since its earliest days, the University has enjoyed a reputation for excellence in the sciences, which as Dr. Cecil points out in his Introduction has been called "the mother of technology." With the founding of the new School of Engineering and Computer Science, and with the continued development of cooperative ties with business organizations involved in highly advanced technology, it is now more important than ever for the University to explore the moral values that undergird progress in and the uses of technology.

On behalf of The University of Texas at Dallas, I wish to express our gratitude to the Chancellor of The University of Texas System, Dr. Hans Mark, to Professor Tom L. Beauchamp, to Mr. Jesse P. Luton, Jr., to Professor Martin E. Marty, and to Dr. Cecil for their willingness to share their ideas and for the outstanding lectures that are preserved in these proceedings.

U.T. Dallas also wishes to express its appreciation to all those who have helped make this program an important part of the life of the University, especially the contributors to the program. Through their support, these donors enable us to continue this important project and to publish the proceedings of the series, thus assuring a wide and permanent audience for the ideas they contain.

I am sure that everyone who reads *Traditional Moral Values in the Age of Technology*, the Andrew R. Cecil Lectures on Moral Values in a Free Society Volume VIII, will be stimulated by the ideas discussed in the five lectures it preserves.

ROBERT H. RUTFORD, President
The University of Texas at Dallas
February, 1987

CONTENTS

INTRODUCTION

by

Andrew R. Cecil

Jean-Jacques Rousseau's first major publication, the *Discourse on the Arts and Sciences*, was written in response to a contest offered in 1750 by the Academy of Dijon for the best answer to the question, "Has the progress of the arts and sciences tended to the purification or to the corruption of morality?" Throughout history, societies have asked the same question, expressed perhaps in different words, when faced with the consequences of such major developments as the invention of the printing press, the steam engine, the electric light bulb, and the automobile, or with the transition from a largely agricultural society to a largely industrial one.

As we face a period of great technological breakthroughs, The University of Texas at Dallas has selected for its eighth annual series of Lectures on Moral Values in a Free Society a theme similar to the question raised by the Academy of Dijon almost two hundred and fifty years ago. The theme is similar because science is the mother of technology, and technology as well as the arts and sciences springs from the culture that helped to create it and is embedded in the culture that nurtures it. Art and literature bring into focus the moral values that act as a unifying force

11

in the life of a society. Science and technology, to
succeed in enhancing human progress, must reconcile
their values with the moral values indispensable for
social cohesion and spiritual progress.

Without such a reconciliation, the only standard
that technology can follow in its development is
technical feasibility within the wide range of scientific
possibilities, without regard to the moral consequences of such development. Without a moral
agenda, the apparent triumph of rationality that
technology promises may turn instead into a major
force for dehumanization. The revolutionary changes
brought about by the computer and by advanced
telecommunications systems, for instance, could become a threat to man's individuality, personal privacy,
and freedom if adequate safeguards are not erected.
When not linked to lofty human values, a mania for
"technology for technology's sake" could bring into
our lives a dull sameness and uniformity, along with
the dangers described by George Bernard Shaw in his
Man and Superman: "I tell you that in the arts of life
man invents nothing; but in the arts of death he
outdoes Nature herself, and produces by chemistry
and machinery all the slaughter of plague, pestilence,
and famine. . . . Man measures his strength by his
destructiveness."

Many aspects of technological change tend to breed
such fears. One fifth of the world's efforts in science
and technology, for instance, is aimed at military expenditures. We depend on technology to defend us,
and nuclear weapons are the principal armament of
modern man. It is estimated that the modern arsenal

of multimegaton bombs has five thousand times the destructive power of all the weapons used in World War II. Such power in itself provokes fears and anxieties and leads us to ask questions. Are the stockpiles of fearsome weapons with the potential to end all life on our planet in harmony with the values necessary for holding society together? Could the assumption that technology could prove a cure for war in fact turn the world into a victim of misplaced confidence?

Such fears and anxieties are not limited to concerns about military technology. There is also much debate about the sorts of cultural, economic, and political changes that will be faced by our technology-intensive society—called by the French the "telematique" society, reflecting the changes in telecommunications and automation—with the further development of computers and communication systems. Will these developments dehumanize the work environment and bring about more centralized (and potentially totalitarian) control over our lives, or will they serve as a decentralizing and liberating force in our society?

Technology need not be a source of alarm and disquiet. While technology cannot solve all the problems that face our society, it can help to solve some of them, including problems that it itself has created, such as high-technology pollution control. When its aim is to unravel the secrets of nature and turn them to the benefit of mankind, technology can even expand the possibilities of the human spirit and enlarge the notion of a more effective form of democracy. Such motivation prevents technology from being pursued solely for the sake of economic gain, with no concern for the common good, or for the sake of usurping power in order to dominate other people.

In his lecture "The *Challenger* and Chernobyl: Lessons and Reflections," Dr. Hans Mark discusses at length the reactions many citizens have to advanced technologies, focusing on the lessons to be learned from the two highly publicized accidents that destroyed an American Space Shuttle and a Soviet nuclear reactor. Dr. Mark explores the kind of human errors that led to these disasters, and contrasts the attitudes in the U.S. and the U.S.S.R. toward technology and especially toward the dangers inherent in some kinds of technological development.

A primary difference between the two accidents and the ways in which they were handled is that from the beginning the *Challenger* disaster and the inquiries into it were open and public. Although the Chernobyl disaster, as Dr. Mark points out, was handled with much more openness and publicity than is usual in Soviet society, there was still a characteristic delay in admitting that any problem existed and a characteristic refusal to admit to all the facts of the case, even those that were obvious to the world through scientific observation and measurement outside the Soviet Union. Despite the setbacks to technological progress and to public trust in technology that these incidents have caused, Dr. Mark predicts that good will result from them in the long run. And—at least within a free society such as that of the United States—open scrutiny and the curbs placed upon risk-taking that are imposed by public opinion ensure that more and more adequate safeguards will be found to protect the safety and health of those affected by technological advances.

Although advances in the field of nuclear power often provoke fear and misgivings, advances in the field of medical technology are more typically greeted with joy and acclaim. Even so, advances in the medical field create moral dilemmas almost without parallel in other disciplines, since medicine deals with human life itself. New complexities of moral and legal thought must deal with new insights in molecular biology and such revolutionary new technologies as the insertion of a baboon's heart into the body of "Baby Fay," the commonplace transplant of human organs, the creation of an artificial human heart, and the widespread use of respirators to prolong life. Alongside techniques that increase our life spans in ways that would formerly have been thought artificial or even magical, there are the new techniques used in the reproduction of human life, some of which have aroused even more controversy: artificial insemination, "test-tube" babies, and surrogate mothers who "rent" their wombs to carry someone else's fetus until birth. In addition to the challenges to the individual conscience that these technologies create, there is the global concern that the new and spectacular medical advances now being made—by retarding the aging process and extending the human life span—will provide a new crop of social and ecological problems for a mankind living on an already very crowded planet.

Our legal system—and the law itself—has not been well equipped to handle these increasingly complex

medical issues. In order to help make often agonizing-
ly difficult decisions concerning life and death, many
hospitals have established ethics committees. National task forces have also been called together to
help such committees set guidelines for issues such as
when to withdraw life-support systems from terminally ill patients and from those who, crippled in mind
and body, are alive only in the most limited sense.
Someone must establish rules and make decisions as
to whether it is in the patient's best interest to extend
his life when it is empty of joy and full of suffering. In
each case, the question must be answered: Is length
of life more important than quality of life? It implies
no lack of reverence to a sense of the absolute value of
human life to assert that sometimes no life may be a
better choice than the shadow of living.

Professor Tom L. Beauchamp, in his lecture
"Medical Ethics in the Age of Technology," points out
that this is only one of the perplexing questions that
modern medical technology raises for the individual
and for society. He outlines the revolutionary change
in outlook that has occurred in recent decades in the
shift from a medical ethic that focuses on the physician and his responsibility to provide healing benefits,
to an ethic that focuses on the patient and his right to
make or at least be consulted in decisions that affect
his life. This shift has been dramatized by technologies that enable us to extend life almost indefinitely,
at whatever cost to the quality of life of the patient.
Professor Beauchamp points out also that the elabora-
tion of technology, with its concomitant great increase
in costs, raises serious questions as to who should
receive its benefits. Many new treatments can be

used only in a limited number of cases during their years of development, and many new treatments involve costly procedures.

Professor Beauchamp sees a new wave of moral issues arising in the near future, as the great breakthroughs that are occurring in genetic research begin to yield fruit in terms of practical treatment. Some of the problems stem from concerns about what information is proper to reveal to those who may be discovered to have inherited genetic deficiencies and to those who are related to them. Other problems will arise as to the propriety of making changes in the structure of the genetic inheritance of our species itself. As challenging as recent developments in medicine have been to our traditional ideas of right and wrong, the developments of the future are bound to be even more challenging.

The most dramatic explorations of our technological society have been our forays into the huge distances of outer space and our use of the knowledge we have gained about the tiniest building blocks of matter—in harnessing nuclear energy and in expanding our understanding of the molecular structure of life itself. Technology has also revolutionized our lives in more mundane ways as well, in many aspects of our everyday lives. Guglielmo Marconi first transmitted a radio message across the Atlantic Ocean in 1901. The computer was born in 1946. The ensuing explosions in telecommunications and computer technology permit information pertinent to all aspects of our lives—economic, social, environmental, political, and military—to be collected and moved from country to

country with the speed of light. As we have seen in
the case of the Chernobyl accident, even totalitarian
countries—that in their fear of losing control of their
people try to build electronic barriers around their
borders—cannot effectively stop the flow of information to and from all parts of the globe. As developments in communications continue to shrink the
planet, will the totalitarian countries open their
societies to the freedoms indispensable to the pursuit
of the fruits of technology, or will they try to tighten
the monopoly they maintain over free communication? Will a political system that keeps such ordinary
devices as photocopiers and mimeograph machines
under strict regulation and surveillance in order to
control what people read and see find it possible to
resist the forces of freedom and of the information
revolution?

Any technology may serve as a weapon of decency
or as a weapon of repression. This is as true of the information storage and retrieval aspects of the
computer revolution as it is of the communications
revolution. Personal information about each of us is
contained in the dossiers of credit institutions, government agencies, religious groups, and political
parties. In totalitarian countries, such information can
be used to control the thought and behavior of the
members of society, to assess their political loyalty,
and to intimidate any dissidents. In the Free World,
we must ensure that these technological changes
permit man to maintain his identity and privacy. We
must determine ourselves whether these changes
demean man's individuality by replacing him with

machines or permit him to continue to thrive on a free flow of ideas that breed progress and a better future.

There are long-term moral implications as new technologies move into the business and professional world. The shift from an economy driven by heavy industry to one driven by information is creating problems of displaced workers, of labor mobility between old and new jobs, and of the need to retrain workers whose jobs have been eliminated by technological change. One of the challenges to business ethics in the age of technology will be finding the proper balance between the self-interest of the entrepreneur, who must search for improvements in productivity, and the well-being of the workers and society in general.

Innovations in technology will assist the businessman who gives proper consideration to long-term prospects for the future by providing him with new insight into possiblities for productivity and for customer satisfaction. Technology can give him a new vision of how to achieve such business goals as high-quality products, new markets, and reasonable profits. Armed with accurate projections of population and mobility and with other statistical data on planned highways, schools, and community systems, the businessman can better evaluate the strengths and weaknesses, capabilities and limitations of his endeavor. Technology will call for more and better-informed participation by the businessman in the decision-making process, not less.

The information-processing procedures that are

making a great impact on the world of business include computerized accounting and computer-based inventory and production-control systems. Computers also now assist in the design, testing, and documentation of new products. CAD (computer-aided design) has been followed by CAM (computer-aided manufacture), wherein a central computer controls the selection of production equipment, guides specific manufacturing processes, and signals maintenance requirements on machinery used by industry. Such developments mean large gains in productivity for industry, but there may be human costs involved in their implementation. Innovations such as unmanned transportation or robotics performing tasks demanding a high degree of skill and accuracy, for instance, could eliminate the need for highly specialized skills among workers. How will our economy absorb the displaced and lessen the hostilities of laid-off workers who will not be rehired into the same fields regardless of any recovery in the economy?

The questions that have to be answered are: Will the pivotal role that technology plays in the economy of the United States solve the problem of declining employment by increasing the rate of job creation in a way that will match the increasing rate of the use of technology? Will the private sector or the government assume the responsibility of retraining workers displaced by technological change? Will technology—by lowering the cost of production and replacing required assets—help to revitalize our basic industries so that they regain the ability to compete on a world-wide basis?

These are not the only moral issues faced by business in our increasingly complex world. Mr. Jesse P. Luton, Jr., in his address "Professional and Business Ethical and Moral Values in the Age of Technology," describes the questionable ethical nature of many practices now commonly accepted in the world of corporate business. The business pages of the newspapers are filled with such new terminology as "greenmail," "golden parachutes," and management buy outs, and Mr. Luton argues that these terms stand for practices that, though now commonly accepted, may go against the best interests of the shareholders of corporations and of the economy in general.

The complexity of modern business and professional life has added new issues to be considered by those who are responsible for making the decisions that affect the lives of employees and of the public. But, as Mr. Luton points out, the same moral principles used to decide the proper course of action still apply today as they did in the past. The traditional values of honesty, integrity, and attention to the rights and needs of others are still the best guidelines for making decisions on a moral or ethical basis in the world of business and the professions.

The eternal verities—as exemplified by the Golden Rule or the Confucian principle of reciprocity—do not change. They are as applicable today as they have ever been. Yet, as Professor Martin E. Marty argues in his lecture "The Many Faces of Technology, The Many Voices of Tradition," the interpretation and expression of these verities do change over time, and, similarly, traditional habits and patterns of thought and behavior are changed by the constant exposure to new ideas and values that our modern transportation and communications systems bring about.

Professor Marty does not see the tension between traditional values and technology as a bad thing. Rather, the challenge of new modes of thinking to traditional ideas can be good in that the challenge forces us to examine our values and strengthen them. It becomes necessary in a pluralistic society such as our own to protect one's values by finding others who share them and by banding together with them. Our glorious American tradition of freedom of assembly has generated an unparalleled number of groups that have been formed to defend a shared set of values, not only churches and political parties but fraternal organizations, trade unions, professional societies, and many other kinds of groups that grow up around shared ideals and visions. The task for those who lead such groups is not only to advance the goals and protect the values of the groups, but to find the common ground that unites all members of our society in bonds of shared tradition and good will.

The use of the media by religious groups poses special problems both for the churches that use them and for the American political process. As we are all aware, the technological revolution has transformed the nature of American political campaigns. President Franklin D. Roosevelt's famous "fireside chats" marked the effective beginning of direct communication between American political leaders and the electorate. In the nineteenth century, public oratory had dominated political campaigns. The famous series of debates between Abraham Lincoln and Stephen A. Douglas remains a well-remembered part of our political history, for instance. In the age of television, this new medium has become a powerful channel of communication by offering a new intimacy between political candidates and voters. Whether in individual appearances or in joint appearances or debates, the candidates are able to employ new means of persuasion to influence the masses of viewers and listeners, thus expanding the nature of our participatory democracy.

The direct involvement of the churches and ordained ministers into this process, however, raises new questions. Television technology offers a unique forum for the crusades of various religious groups, especially for evangelicals and fundamentalists. Many of these are not content to try to mobilize the consciences of their millions of viewers by drawing on the heritage of their movements and on their belief in the word-for-word accuracy of the Bible. Rather, along with spiritual teaching, the religious broadcasters tend more and more to convey a call for social and political action. There is a danger that, by thus entering the political arena, the churches will undermine their effectiveness as moral guides. Throughout the history of our nation, religion has provided an essential foundation for our system of democracy, and the churches have never remained silent on issues with moral content. The increased influence, however, of religious television programs and especially of

preachers who have become political celebrities has
introduced the peril of the entry of the churches into
the game of practical politics. This development has
provoked a justifiable fear that political involvement
will cloud the main responsibility of our churches,
which is to concentrate on moral goals and to retain
their objectivity, free of the taint of political partisan-
ship.

The prominent lecturers whose addresses appear in
this book have raised many of the questions posed by
technological development for the future of our
society. Their views are certainly more encouraging
than the defiantly negative answer Rousseau gave in
response to the question asked by the Academy of
Dijon—whether the arts and sciences tended to the
purification or to the corruption of morality. As the
popularizer of the idea of the "noble savage,"
Rousseau distrusted the idea of progress. The myth of
a natural man uncorrupted by the advances of civilization was so enticing to him that he could see only
corruption in the advent of new developments in the
arts and sciences.

The participants in the 1986 Lectures on Moral
Values in a Free Society, while admitting that ours
will be a different society in the years ahead, see our
defense against the possible abuses of technological
developments in the traditional moral values of our
culture, which stress the primacy of human sensitivity
and affection, and in our continuing quest for
knowledge and the development of our intellects.
Their expectations remind us of some observations of

John Adams in a letter of October 29, 1775, to his wife, Abigail:

> "Human nature with all its infirmities and depriva-
> tion is still capable of great things. It is capable of
> attaining to degrees of wisdom and goodness
> which, we have reason to believe, appear respect-
> able in the estimation of superior intelligence.
> Education makes a greater difference between
> man and man than nature has made between man
> and brute. The virtues and powers to which men
> may be trained by early education and constant
> discipline are truly sublime and astonishing."

As we move from an industrial to an informational society, technology will affect what we teach and the way we teach it. Computers, for instance, will have a vital role in the educational process. (The National Commission on Excellence in Education appointed by U.S. Secretary of Education Terrel H. Bell recommended in its report, *A Nation at Risk: The Imperative for Educational Reform*, that every student receiving a high-school diploma should have at least one-half year of computer science.) We undoubtedly need technology. And we need citizens trained in the uses of technology. Schools must make full use of instructional technology, but the need for technological literacy does not mean that we can afford to neglect the disciplines that nurture spiritual growth. Research should not replace teaching as the main purpose of our institutions of higher learning, nor should television replace the teacher in the classroom. In the realization that new frontiers lie ahead, we must forge an educated society by rediscovering our humanistic and spiritual roots and by learning how to apply their principles to the situations brought about by technological change. Such subjects as literature, history, the fine arts, philosophy, and foreign languages may not produce results identifiable by digital enumeration or rewarded by the dollar sign, but they still deserve the place of honor in our curricula.

While technological progress calls for more emphasis on the study of science and mathematics, these disciplines are not self-contained—they interact with the study of the humanities. The need for personnel trained in technology should not obscure the truth that the world does not exist in compartments; its communities are interrelated. A free society has a multifaceted quality, and each facet affects the others. Our society is comprised not only of technocrats, manufacturers, producers, and consumers, but also of philosophers, writers, poets, and artists who widen our horizons, teach us to think clearly, and give us an understanding of the meaning of communications skills. In this age of information, effective communication is a primary vehicle for the success of businesses that use technology.

In my lecture "The Unchanging Spirit of Freedom," I have tried to point out that in the age of technology, just as in other periods of the history of mankind, the liberating virtue of freedom is a reality that must be recognized. Nothing can replace the spirit of freedom. No scientific discovery, no technological change can weaken man's hunger for freedom; it has proved to be the unchanging force which underlies all that men have undertaken and accomplished.

In fact, freedom of thought is necessary for the full flowering of the spirit of innovation that results in progress, and freedom—along with the other traditional moral values upon which this country was founded—will continue to exercise its formative influence on human behavior in the age of technology. The record of our nation, its ingenuity and its resilience, demonstrates that we will succeed in being wise managers of change.

In the 1986 Lectures on Moral Values in a Free Society, the theme is repeatedly emphasized that the technological revolution should not generate fear but, rather, hope for a better world—for an affluence ample enough to allow time for discovering the important things of life, for permitting us to spend more time in cultural and educational activities, and for unraveling the secrets of life and of the eternal, seamless web that is man and nature. Revolutionary new technologies may raise new ethical issues and prompt changes in the interpretation of traditional values, but as long as those who use them recognize the importance of the values that hold a society together, they will contribute to the advances of civilization and the improvement of the standards by which we measure the values of human life.

THE *CHALLENGER* AND CHERNOBYL: LESSONS AND REFLECTIONS

by

Hans Mark

Hans Mark

Hans Mark was named Chancellor of The University of Texas System, the fifth-ranking university system in the United States in enrollment, on May 30, 1984.

Before joining the U.T. System, the Chancellor was Deputy Administrator of NASA, which directs the nation's space program. From 1977 to 1979, he was undersecretary of the U.S. Air Force, and was named Secretary of the Air Force by President Jimmy Carter in 1979. From 1969 to 1977, he was director of the NASA-Ames Research Center at Moffett Field, California.

Chancellor Mark, whose academic and scientific career has spanned more than three decades, is a Fellow of the American Physical Society, the American Institute of Aeronautics and Astronautics, and the American Association for the Advancement of Science. He played a pioneering role in the study of x-rays from stars and did early work with what were later identified as black holes, and is author and co-author of more than 100 scholarly articles.

Dr. Mark held faculty appointments at Boston University, Massachusetts Institute of Technology, and Stanford University and served as Professor and Department Chairman at The University of California Berkeley campus between 1960 and 1969. He obtained an A.B. degree in physics from The University of California at Berkeley in 1951 and a Ph.D. in physics from MIT in 1954. He is a member of Tau Beta Pi, Sigma Xi, Phi Beta Kappa, and the National Academy of Engineering. NASA awarded him the distinguished Service Medal in 1972 and 1977, and he received the Decoration for Exceptional Civilian Service of the U.S. Air Force in 1979 and the Distinguished Public Service Medal from the Department of Defense in 1981. He is the recipient of two honorary degrees, the honorary Doctor of Science degree from Florida Institute of Technology in 1978 and the honorary Doctor of Engineering degree from Polytechnic Institute of New York in 1982.

THE *CHALLENGER* AND CHERNOBYL: LESSONS AND REFLECTIONS

by

Hans Mark

"You mock the very skill that proves me great."
Oedipus Tyrannus

The first six months of 1986 were not good ones for those of us who have an interest in the development of new technology. Two accidents, the loss of the Space Shuttle *Challenger* in January and the destruction of Unit #4 at the Chernobyl nuclear power station in the Ukraine in April, have raised serious questions about our ability to manage high technology. Though vastly different technologies are involved in each case and though the accidents happened in two different nations that have very different attitudes toward both management and public information, the two events have one important thing in common: in each case, a large-scale system with great technical complexity failed catastrophically.

Such failures give us pause. It seems an essential human activity to want to learn from our mistakes, and in both these cases, we can, I think, learn a great deal that will contribute to the health of our society and to the success of its future technological

31

endeavors. But more importantly, such accidents call our attention once again to the existence of forgotten truths concerning the human condition. As human beings, we fail, but our very humanity depends on our accepting the risks and responsibilities for failure and of choosing to go on. To renew our faith in the values that encourage us to discover new seas and sail upon them regardless of the risks is, I believe, the most important lesson *Challenger* and Chernobyl have to offer. It is not an easy lesson: it is one our deep-seated fears of risk and of freedom tell us to ignore. To do so, however, would be to sacrifice not only our technological prowess but the values of openness, responsibility, and rationality upon which our society, indeed our very humanity, depends.

Let me first narrate what happened and explore, to the extent presently possible, why it did so. I have been, as it turns out, closely associated with the technologies concerned in both accidents. I spent a decade teaching nuclear engineering at the University of California in Berkeley and for five years of that period, I was the supervisor of the University of California's Berkeley Research Reactor. I therefore know something about nuclear reactors and what precautions must be taken to operate them safely. I also spent fifteen years closely associated with the nation's space program. During much of that time, I was directly involved, first in the development of the technology for the Space Shuttle and later on in working out the operational procedures for flying the shuttle. Thus, I may have a unique perspective from which to draw lessons from both of these tragic events.

I

Destruction came to the Space Shuttle *Challenger* seventy-three seconds after it was launched from the Kennedy Space Center in Florida on January 28, 1986. The proximate cause of the accident was a failure in the seal of the joint between the lowest and next to the lowest segment in one of the Solid Rocket Motors (SRMs) of the vehicle. (This joint between the segments is called a "field joint" because it is made up in the "field," that is, at the Kennedy Space Center rather than at the manufacturer's plant.) That much has been established beyond any doubt. As with most technological failures, behind the principal cause lie many others. In this case a combination of factors, starting with the O-rings, ultimately led to the accident. As early as 1977—four years before *Columbia's* first flight—engineers knowledgeable in the area of seals and joints had raised questions regarding the design of the O-rings. (These objections are documented in the Report of the Presidential Commission chaired by former Secretary of State William Rogers that investigated the *Challenger* accident. See pages 123-124.) At the time, management at the National Aeronautics and Space Administration decided that these objections were not serious enough to warrant changes and retained the design. I was not in NASA in 1977 and cannot, therefore, speak from personal experience as to why NASA management chose not to heed these warnings. What is clear is that the seed for the *Challenger* tragedy had been sown nine years before.

My own part in the chain of events that led to the

accident began when I became Deputy Administrator
of NASA early in 1981. At that time, our engineers
began to question whether these "field joints" on the
SRM with their O-ring seals were really fail-safe. During the design of the Space Shuttle, NASA tried to
make as many of the subsystems as possible "fail-
safe"; that is, to design them in such a way that a
single-point failure would not have catastrophic consequences. In the case of the "field joints," this was
accomplished by putting two O-rings in the joint on
the theory that if the first one failed, the second
would do the job.

Questions as to whether the double O-ring system
was really fail-safe continued to be raised during 1982.
Nevertheless, in February or March of 1983, Mr. L.
Michael Weeks, the Deputy Associate Administrator
of NASA for Space Flight, signed a memorandum
waiving the fail-safe requirements for the field joints
in the SRM. I remember discussing that matter with
him at the time and concluding that such a step was
justified. I argued that we had more than a hundred
successful firings of the Titan Solid Rocket Motor with
a seal of somewhat similar design containing only one
O-ring. I thought because of the Titan precedent that
the risk of failure was small. I am not sure now that
this judgment was correct.

We did, however, have other serious difficulties
with the shuttle vehicle that demanded much more of
my time than did the O-ring seals. There was, for example, the problem with the erosion of the SRM
nozzle. On one flight we came within a few seconds of
burning out the nozzle before the rocket fuel ex-

hausted itself. As a result of skirting disaster, we conducted a series of very detailed reviews to get to the bottom of this problem and then to fix it. We succeeded in attributing the problem to the failure of a third-level contractor to follow the proper procedure in processing the heat-resistant synthetic material that lines the nozzle. We had difficulties with hydrogen leaks in the aft end of the vehicle. I remember that we postponed the first flight of *Challenger* for more than a month looking for hydrogen leaks in the aft engine compartment. We noticed this problem for the first time during the flight readiness firing of the *Challenger's* engines. We had problems with the auxiliary power units on the orbiter. These tended to run hot, and during every flight we feared power failures. At the time, all of these things were more important in my mind than the problem with the O-ring seals.

The O-ring seal problem did again command my attention just before I left NASA in 1984. On the tenth flight (STS-41B) we noticed some charring of the O-rings in the lower field joint. This phenomenon had been observed once before on the second flight (STS-2), but when it did not reappear, we thought it a one-time event. When we saw it again on the tenth flight, the question of what should be done was discussed at the Flight Readiness Review for the eleventh flight (STS-41C). After the completion of the Flight Readiness Review, I issued an "Action Item" asking for a complete review of all the SRM seals and joints. My intention at the time was to review this problem in the same manner as we had done with the SRM nozzles when we had problems with erosion.

Unfortunately, this review was never held. I made the
decision to leave NASA about two weeks after signing
the "Action Item," so the matter was apparently
dropped. (The "Action Item" was signed on March
30, 1984, and I made the decision to leave NASA in
mid-April 1984.) The due date for the review was May
30, 1984, and by that time I was a lame duck. I should
have insisted on holding the review anyway. Perhaps
then more attention would have been paid to the
problem once I had left.

The subsequent history is explained on page 132 of
the Rogers Commission report. The people at
Marshall Space Flight Center and Thiokol, the contractor, decided that they would develop a plan to fix
the O-ring problem rather than review the matter
with the higher level NASA management. It is
apparently for this reason that nothing was done for
fifteen months to make NASA management at the
Headquarters level more aware of the problems that
were developing with the O-ring seals at the time. It
is also for this reason, very probably, that former
NASA Administrator James M. Beggs and other high-
level officials at NASA could claim that they were not
aware of any really serious problems with the O-ring
seals. (See page 135 of the Rogers Commission
report.) A complete review of the O-ring seal
problems on the SRM was finally held at NASA
Headquarters on August 19, 1985, fifteen months
after the original request was made. Even then, the
most senior person who attended the review was Mr.
Weeks, a Deputy Associate Administrator of NASA.
Neither the Administrator of NASA nor the Associate
Administrator for Space Flight was present.

To summarize, I was aware of the problem with the O-ring seals in 1983 and 1984. I felt that it would have to be dealt with as a result of the review I called for in a manner similar to the reviews we had for other problems we experienced. I did not think, however, that the O-ring seal problem was as serious, for example, as the SRM nozzle erosion problem or some of the problems we were having with the shuttle main engines. Nevertheless, I felt that the time had come to investigate closely the effectiveness of the O-rings.

I have asked myself over and over again, whether I would have flown on that January 28. I was involved in the launch decision for twelve shuttle flights, so I do have some experience in how these things work. I remember that prior to the launch decision in every flight there was always at least one group of subsystem engineers with a problem that caused them to advise us not to launch. Sometimes we took their advice and postponed the launch; at other times, we went ahead and flew anyway. Simply because a group of engineers opposed the launch out of the fear a subsystem might fail was not enough to cancel it. In view of my experience, I do not know whether the recommendation of the Thiokol engineers not to fly would have been enough to persuade me not to launch the *Challenger* carrying Mission Commander Dick Scobee and his fellow astronauts.

I do have to confess that when I saw the pictures of the ice on the launch pad in the Rogers Commission report, I was surprised that the NASA management gave the go-ahead to fly. The launch pad structures and the gantry were both completely covered with ice and there were a great many icicles. Icicles become missiles when the pad vibrates during takeoff, and they can easily damage the vehicle. I had always been especially concerned about damage to some of the tiles of the thermal protection system. The tubes that carry the liquid hydrogen for the regenerative cooling of the shuttle's main-engine nozzles are also vulnerable to damage by flying ice. Obviously, a rupture of these tubes could lead to catastrophe. During the years I had anything to say about whether or not to launch, we cancelled launches with much less ice on the pad than I saw in those pictures.

These are my impressions, but what I have said with respect to the decision to launch is clearly second guessing. I was not there; I will never know what I would have done. Clearly, there was a failure in communication in the decision to launch as well as in the design and in the flight readiness review process. The necessary information on which decisions should have been based apparently never reached top-level NASA management. In the case of the danger signals that we had of O-ring failures, the information did reach the top management level in March 1984 during the Flight Readiness Review for the eleventh launch (STS-41C). However, the review of the situation that I requested subsequent to the Flight Readiness Review that should have been conducted was not held. Thus, the problem was never properly explained to the Administrator and his immediate assistants because people were apparently under the impression that things were being fixed. This happened even though O-ring erosion problems much more serious than the ones experienced on the second and on the tenth flights were observed on flights after the eleventh flight (STS-41C). In the case of the decision to launch *Challenger* on the morning of January 28, 1986, the three senior people who bore the responsibility, the Acting Administrator of NASA, the Associate Administrator for Space Flight, and the Director of the Space Shuttle Program, did not know that the Thiokol engineers had, because of the cold weather, raised objections to flying. This, to me, is evidence that the good communications between top management and the engineers in the field, a factor crucial to the success of any launch, simply did not exist.

It would be easy to conclude that a better information system would have solved the problem—just make sure that the working engineers report everything they are doing and all will be well. The situation, unfortunately, is not that simple. Setting up that kind of an information system would guarantee that much more would be written than read! No, to correct these problems in communication, people at the top of the organization must ask the right questions. It is a two-way street. The people in the ranks must tell the truth and promptly write clear reports. The people at the top must make judgments about which questions to ask, and they must have the proper instincts to ask the right ones. In my judgment, both the working engineers and the NASA leadership failed to meet these responsibilities on that fateful day and during the months that preceded it.

The *Challenger* tragedy was played out in the full glare of all the attention that the public media could provide. Many of us actually saw the accident in "real time" on our television screens. All of the proceedings of the subsequent investigation of the accident by the Rogers Commission have been published. A thorough effort has been made to make certain that everything that can be known about the accident has been made public. This policy of making all information available is traditional with NASA, and while there have been some minor glitches in timing the release of information, the normal policy of informing the public fully has been followed.

II

On April 26, 1986, at about 1:24 in the morning, observers outside the Chernobyl nuclear power plant about seventy-five miles north of Kiev in the Ukraine reported hearing two explosions and seeing burning material and sparks shooting into the night sky. From information provided by the Russians at a conference held by the International Atomic Energy Agency in Vienna, August 25-29, 1986, and from inferences made from the type of nuclear reactor and the dispersion of radioactive material, we can put together a good, if not complete, picture of what happened and why.

Unit #4 at Chernobyl, a RBMK-1000 graphite-moderated uranium reactor of 3200 megawatts thermal energy, produced about 1000 megawatts of electrical energy while running at design power level. The reactor and the other three units in this large plant shared some common subsystems, for example cooling and electrical utilities.

Graphite-moderated reactors are inherently more dangerous than the water-moderated reactors commonly used by commercial power companies in the United States. A graphite-moderated reactor does not have the large negative temperature coefficients characteristic of water-moderated units. This means that unlike water-moderated plants, whose reactivity decreases as the temperature increases, graphite-moderated plants have a reactivity that increases slowly as the temperature increases. Therefore, the reactor is inherently unstable and requires a complex and sensitive control system to manage the reactor power level. No graphite-moderated power reactors are operated in the United States. The Russians use graphite-moderated reactors because they do not require highly enriched uranium to operate. Thus they are less costly but some increased risk is incurred in their operation.

When the accident occurred, the operators of the plant were conducting an experiment: If the reactor should fail, they wanted to know how much energy the inertia of the turbine rotor would contribute to auxiliary electrical supplies before standby diesel generators could be started. The test, which began at 1:00 P.M. on the afternoon of the 25th, called for reducing the reactor's power over a five-minute period from 100 percent to 50 percent, from 3200 megawatts to 1600. Then one of the two turbogenerators was taken off-line, the system it supported being transferred to the other turbogenerator.

At the request of a local electricity dispatcher, the experiment was held up for about nine hours. It was

not until 11:10 P.M. that preparation for the test continued. The plan called for a further reduction in the
reactor's power to 700-1000 megawatts, but the operators apparently failed to activate on time the
automatic controls designed to stabilize the reactor
when operating at lower power levels. Consequently,
the power plummeted to below 30 megawatts.

The operators came dangerously close to losing control of the reactor at this point. In order to increase
the reactor's power, they pulled many of the manual
control rods; this led to an increased reactivity and the
power went up to 200 megawatts. But they were dealing with an increasingly unstable reactor core, and the
prudent course would have been to halt preparations
for the test and to shut down the reactor.

The long delay may have led to an attitude of the
operators of "let's get this test done, procedures be
hanged." But whatever the reason, they plunged
ahead, and despite their cavalier treatment of the
reactor, by 1:22 A.M. they had brought it to a fairly
stable condition. They then committed two crucial
errors, and their luck ran out. In order to run one test
and then repeat it quickly, they shut off the
emergency protection-signal system from the turbine
stop valve, and they reduced the coolant to the initial
level required for the original test conditions.

Reactivity increased, and though the operators
received a printout from the fast reactivity evaluation
program that warned of the need to shut down the
reactor, they went ahead. Two things, happening
simultaneously, were rushing the reactor to disaster:
reactivity was increasing and the reduced cooling was
quickly causing the cooling water to approach the

flash point where it rapidly becomes steam. The state
of the reactor was such that control rods would have
had to go well into the core to provide the neutron
absorption necessary to stabilize the reactor; this
couldn't be done, probably because "melt down" had
begun. It was too late.

Very probably during this rapid power excursion,
one of the fuel elements in the reactor ruptured. The
Chernobyl reactor is fueled by slightly enriched
uranium-oxide pellets enclosed in zirconium-alloy
fuel rods. These rods in turn are kept in zirconium
pressure pipes through which the cooling water flows.
The pipes are called channels, and there are eighteen
fuel rods mounted in each channel. Very probably one
of the channels also ruptured due to overheating during the power excursion after the failure of the fuel
element. This would happen because steam would be
created during the excursion as the cooling water
boiled at one or more places inside the reactor core.
The excessive pressure caused by the local boiling was
apparently sufficient to rupture the zirconium
pressure channel. Above 1000°C, steam reacts ex-
othermically with zirconium to make hydrogen. Once
this process started inside the reactor core, there was
a source of hydrogen and, very probably, the graphite
moderator also started burning since it will burn
slowly in hot steam, producing carbon monoxide.
Carbon monoxide can react explosively with
hydrogen. Since the zirconium-steam reaction is ex-
othermic, it spreads rapidly once it starts and, therefore, a source of hydrogen was created, with the
slower (endothermic) graphite-steam reaction producing the other ingredient, carbon monoxide, necessary

to have an explosion or two explosions, though why there were two explosions remains unclear. It may be that the first one was simply due to very high-pressure steam and that the second one was caused by the mechanism I have postulated.

From the pictures published by the Russians, the explosions completely destroyed the building that houses Unit #4. The explosions also probably destroyed the remaining cooling system and so the core of the reactor melted. The Russians deny a melt down occurred, but the evidence suggests otherwise. We know, for example, that temperatures in excess of 2500°C, a white-hot, melt down temperature, must have existed because isotopes of very refractory materials such as ruthenium 103 and ruthenium 106 were picked up outside the borders of Soviet Russia. It is important to understand that Russian practice is not to put containment buildings around their reactors. In this country, all commercial power reactors are equipped with containment buildings designed to contain chemical explosions that might occur during a reactor accident.

The environmental effects of the Chernobyl accident became apparent a few days after the event when the radioactive cloud that was released into the atmosphere drifted over Sweden. The Swedes saw radiation levels at certain places 100 times above the normal background, and they were also able to analyze the isotopes that were released during the accident. From the measurements that could be made outside of Russia, the best estimate is that about 40 percent of the inventory of volatile radioactive material in the reactor core was released. In the case of iodine 131, which is the most volatile of the radioactive fission products, approximately 80 million curies of this material were dumped into the atmosphere. (A curie is defined roughly as the radioactivity of one gram of radium in equilibrium with its surroundings.) Iodine 131 has a half-life of about eight days and so it was possible to trace back the events of the Chernobyl accident by looking at the distribution and time dependence of iodine 131 activity. The fact that background radiation readings were increased by a factor of 100 in various localities is quite consistent with this number. The background radiation level is about .04 mrem (milli-roentgen equivalent in man) per hour. The background dose in one half-life of iodine 131, that is eight days, is therefore: 8×10^{-3} rem or about 10^{-2} rem. Estimates made in Stockholm were that the population received a total exposure of about one rem during the decay period of the iodine 131. (These figures are taken from an analysis of the radioactive fallout patterns made by the staff at the University of California's Lawrence Livermore National Laboratory.)

There is no question that the release of radioactivity from the Chernobyl plant was the largest that has ever occurred from a nuclear reactor. For comparison, the Three Mile Island accident in 1978 caused the release of approximately one curie of iodine 131. The Chernobyl release was 80 million times larger. Another interesting comparison is with the radioactivity release by a nuclear weapon. One day after the detonation of a Hiroshima-type nuclear-fission bomb (20 kiloton yield) in the atmosphere, there will be about 200 million curies of fission-fragment radioactivity. Thus a very rough estimate is that the fallout produced by the iodine 131 from the Chernobyl accident is roughly one third of that which would be produced by a Hiroshima-type fission bomb. Many assumptions go into this estimate, including fallout patterns. The estimate, however, is probably conservative since it is based on only the release of volatiles such as iodine 131 from the reactor core. If refractory materials are included, then the radioactive fallout from Chernobyl is probably as large or larger than that which would come from a Hiroshima-type nuclear weapon.

The Chernobyl accident has so far (as of late August) resulted in thirty-one deaths. These were operating personnel, killed by the shock effect of the explosion, by the secondary effects such as collapsing buildings, and by the immediate effects of very high levels of radioactivity. Another 170 or so have been hospitalized with acute radiation sickness, and the death rate among them will be high.

About 100,000,000 people in western Russia and eastern Europe were exposed to higher than normal radiation levels. In the next fifty years, estimates are that there will be about 10,000 extra cancer cases due to these radiation exposures. These estimates are based on the so-called "linear" extrapolation which says that genetic damage due to radiation can be extrapolated linearly to zero from the lowest measurable biological damage levels. The same population of about 100,000,000 people will have approximately 20,000,000 cancer deaths from other causes during the same fifty-year period. It is doubtful, therefore, whether the estimated 10,000 extra cases due to the Chernobyl accident will ever be statistically measurable.

In addition to the direct effects that I have mentioned, about 40,000 people have been evacuated from the Chernobyl area. It is likely that the exclusion zone around Unit #4 will have to be maintained for some years. The Russians have excluded people from living in a region with a radius of about eighteen miles centered at the destroyed reactor. Those living within that area have been evacuated; people will be able to move around the exclusion area and even to work in it, but for limited periods of time depending on the industrial radiation exposure rules that the Russians adopt.

The immediate cause of the Chernobyl catastrophe was human error compounded by the pressure to complete a task and to complete it quickly. Valery Alekseevich Legasov, head of the Soviet delegation to the Vienna conference, noted, for example, that the operators would have felt pressure to complete the tests during this shutdown because the next planned maintenance was more than a year away. Moreover, the rules for safe procedure were only formal. Operators were, for example, instructed to ask permission of the plant shift foreman before carrying out experiments. Although safety protections were built into the plant's systems, the operators were able to override them. The operators, whether out of ignorance of the basic physics of the reactor, out of pressure to do the job, or out of both, shut down automatic safety systems to conduct an experiment without proper authorization. That the RBMK lacked diverse shutdown systems allowed human errors to proceed to disaster.

In stark contrast with the *Challenger* accident where the whole drama was played out in full view of the public, the Russians experienced acute embarrassment as a result of the piecemeal way in which the details of what happened at Chernobyl came to public attention. With no tradition of operating openly as in the case with the nuclear industry in the United States, the Russians obviously were improvising. Furthermore, their first instinct was probably to keep as much as possible about the event secret. Though there is much more that we need to know concerning what happened in the Ukraine on that fateful day, that the Russians were finally not able to keep the events secret and that they have been as forthcoming as they have been is perhaps a hopeful sign. For if the conduct of high technology involves risks in an open society, these risks are accentuated in a closed one. This is but one of the lessons of *Challenger* and Chernobyl. It is to this and other lessons that I now wish to turn.

III

Challenger and Chernobyl confirm Samuel Johnson's observation that men more often need to be reminded than taught. Both disasters hold out for us obvious lessons, ones often forgotten but nevertheless vital to our well-being. Aside from information we have gleaned concerning details of engineering, two lessons of a more general nature stand out.

The first is quite simply that ignorance is the father of disaster, a father whose progeny multiply hideously when powerful, complex technologies are involved. Ignorance of two types contributed to these tragedies. First, the engineers in NASA did not really understand how the lower "field joint" on the SRM actually worked. And it is equally clear that responsible engineers at Chernobyl did not understand how the control system of their nuclear reactor worked. Secondly, this ignorance was, in the case of *Challenger* and probably in the case of Chernobyl too, compounded by the ignorance fostered by a flawed decision-making process. It is not enough to know; we must communicate what we know. Ignorance about how people and machines interact as well as about how people interact with people led to both catastrophes. Education is absolutely essential if we are to avoid such disasters in the future.

So too is discipline. By discipline I mean the obligation of working engineers to report honestly their results and of the management to ask the right questions. We do know what happened in the case of the *Challenger* where the discipline that should have been applied during flight readiness reviews broke down. In the case of the Chernobyl accident, though we do not know exactly what happened, it is likely that the same breakdown in discipline occurred, the breakdown perhaps being much worse. There is some evidence, for example, that the leadership in Moscow did not know what was happening until radioactive debris actually was detected by people outside the borders of Soviet Russia. Very probably they were told that there had been an accident, but the seriousness of the situation was withheld from them by the people at the Chernobyl site. That they waited for three days to carry out the massive evacuation around the Chernobyl area suggests such a conclusion and points to a massive failure in discipline.

That we need to know and that we need the discipline to use what we know are hardly startling truths, and why we require disasters to return us to what is readily available through common sense remains a mystery. But if we take these truths to heart, we will, I believe, fly the Space Shuttle again. My guess is that it will be within two years. The Space Shuttle program will go on roughly as planned. There will be a delay of one and one half or two years, but then flights will go on, and they will be better and safer than they were in the past. I draw this conclusion because it is in accord with history. The Apollo-204 fire in 1967 caused the same kind of situation. We experienced a delay in the flight program. Following a thorough investigation, we improved both the technology and the management of it. These improvements gave us confidence to take some risks by shortening the planned test-flight program, but these risks, built upon confidence in our information, allowed us to meet the objective set by President Kennedy to put a man on the moon by 1970 and to bring him back safely. We will, I believe, take such informed risks again. As a result, the Space Shuttle will fly, we will build the Space Station as planned, and the benefits promised by space exploration will become tomorrow's realities.

Learning the lessons from Chernobyl could lead to an equally optimistic future for the nuclear power industry and for research into nuclear energy both here and perhaps in Soviet Russia as well. Though this may seem paradoxical to some, the Chernobyl accident, I believe, will make it easier to build nuclear-power reactors, in large part because the destruction of Unit #4 at Chernobyl was the "maximum credible accident" that the Nuclear Regulatory Commission requires people who operate reactors to analyze. Heretofore, we have never had one. The accident at Three Mile Island has been, up to now, the only existing standard by which to judge what a "maximum credible accident" really entails. It caused a release of iodine 131 that was almost 100 million times smaller, and this was by far the biggest release of radioactivity that we have ever had in the American commercial nuclear power reactor business. Thus, the opponents of nuclear power have had builders and operators over a barrel. They could invent any and all terrible consequences of a "maximum credible accident" without having to be concerned about what might happen in the real world.

This will not be true from now on because the effect of the Chernobyl accident will eventually become public knowledge. The event at Chernobyl is, without question, a "maximum credible accident." Thus we now have a data base from which much better judgments than we have made in the past can be developed. What will be the consequences of the

Chernobyl accident when all the facts are known? About two hundred people will die of the immediate exposure to large doses of radiation. The prompt effect of the accident is thus similar, in terms of casualties, to a major airliner crash. An area of about eighteen miles in radius surrounding the plant has been evacuated and about 100,000 people have lost their homes. It is likely that many of these people will be able to return in a few years' time once appropriate decontamination procedures are complete. Finally, there will be no detectable increases in the number of cancer cases in the region although there will be a slight theoretical increase. In making these observations, I certainly do not mean to take lightly the very real destruction, suffering, displacement, and psychological fear caused by the Chernobyl accident. However, confronting this situation openly, the possibility exists that the public perception of nuclear reactor problems will become more accurate and more reasonable.

If such are the results of the *Challenger* and Chernobyl disasters, we will reap further insights and useful technologies from both our exploration of space and of nuclear energy. And such results will affirm the values of openness, of confronting failure, and of taking responsibility for it. When we in the United States have had such accidents in the past, we have, as we did with *Challenger,* opened them up to often excruciatingly painful public exposure. But the result has been a catharsis, a healing, and a renewal of strength. The relative openness of the Russians toward the Chernobyl accident, as shown by their report to the International Atomic Energy Agency, is a positive sign. It is possible that they, if only in this one case, have begun to recognize that the price for not being open, for not taking responsibility, is a cultural neurosis that not only cripples science and engineering but absorbs the creative energies of an entire culture. The values of openness, responsibility, and rationality are inimical to a totalitarian society. The tendency of such a society is to treat error not as an opportunity to learn but as a sign of weakness, something to be hidden rather than as something to be explored as a source of future strength. We can only hope that the investigation that has gone into Chernobyl will become the rule not the exception in Russian life and that as a culture they go on, as we have often done in our past, to confirm an intuition of James Joyce that "[people] of genius make no mistakes. [Their] errors are volitional and are the portals to discovery."

These disasters, particularly Chernobyl, have within them, however, the power to provoke a far different response. "Four months later," a *New York Times* reporter tells us, "Chernobyl still has a raw and frightening actuality in the European consciousness." We are receiving reports from Russia of the conscripted labor being used to clean up the site, of their exposure to radiation, and of their fears of sterilization and worse. Will the Russians admit they are doing this? In Britain, the government is destroying affected sheep by the thousands, and Sweden may have to destroy an estimated 100,000 reindeer who have eaten radioactive lichen in Lapland, a destruction that threatens the way of life of 15,000 Laplanders. The Swedish Prime Minister, Ingvar Carlsson, recently lamented that "the Chernobyl accident has spread radioactive iodine and cesium over our fields, forests, moors, and lakes." And he went on to conclude that "nuclear power must be got rid of."

Many in Europe and in the United States no doubt share the Prime Minister's sentiment. For truly, it is a fearful thing to confront the unleashed powers of the universe. But the danger for all of us in this reaction is that nuclear power, its risks and rewards, will not be weighed in a rational manner; instead our irrational fears will exile it to some terra incognita of the mind with the hope of forgetting it. In other words, what will have decided the fate of nuclear power is not our best information, openly and rationally examined, but the passions erupting out of the dark regions of our being.

That this is a distinct possibility should come as no surprise. At their best, the thoughts of most people have never been terribly clear about technology, and I imagine that is so because they usually come inter-mingled with thoughts of death and a dread of freedom, a sense that we do not know where we are going. Both of these emotions are, as the protests against nuclear power testify, very strong, perhaps among the passions the two most powerful.

Our fears and passions do, of course, have an im-portant role to play in making decisions. They serve as warnings, as intuitions of trouble. And we best heed them. At the same time, the passions alone should not be the basis for decisions. There are, for example, good arguments to be made against manned space flight, in-deed given the pressing social problems here on earth, good arguments for not spending billions on space ex-ploration at all. I think these arguments are wrong, that the exploration of space will, as it has in the past, contribute enormously to man and to the planet he inhabits. Rational men, conducting rational, open debate, can, however, come to different conclusions on this matter. But should we stop our ventures into space out of fear of risk—a fear really of freedom and of death—then we will have sacrificed more than the space program. We will have given up what makes us human: the capacity to confront, weigh, and finally to act despite our fears.

As I said, I do not think the space program will be grounded for long. That it will not, that some 89 percent of the American people surveyed in a recent poll said that the shuttle should go on despite the risks, affirms certain basic values in our society. To face a tragedy openly; to investigate it thoroughly regardless of the consequences; to discover some of the reasons for the failure; to conclude that though man can overcome some of the problems, failure will inevitably come again; to confront all of this, painful as it has been, and to go on is testimony to the values of an open, rational, and finally courageous society.

But as the space program and the *Challenger* episode reveal our openness to light, Chernobyl and the general reaction worldwide to nuclear power may reveal our susceptibility to the forces of darkness. While working on this paper, I asked several people who work in our office about their reaction to nuclear power. One expressed a concern about the disposal of nuclear waste, but another simply said: "I hate it. I don't want to glow in the dark. A friend of mine has a microwave, and he nukes everything." Another, when told of the subject matter of the paper, said cheerily: "I hope you don't begin to glow." These are

somewhat humorous manifestations of a deep irrational response to nuclear power, the death masks
that inevitably accompany antinuclear demonstrations
being a more sober example. These fears need to be
understood and ultimately they must be defused.

If we do overcome the present fear of nuclear
power, if we do go on to explore its potential to make
better the life of man on earth, it may well be because
the Russians, albeit uncharacteristically, have been as
open as they have about Chernobyl. Attempts to
choke off information, to remain secretive, would
have led, and did for a time, to greater fear and a
predictably dimmer future for nuclear energy.

In *Challenger* and Chernobyl we have, as we so
often do in Western culture, rehearsed once again the
action of Sophoclean tragedy. The Greeks of Oedipus'
Thebes need to halt the plague destroying their city.
Some are truly ignorant of the cause; others seek to
keep it hidden. Oedipus alone cannot "leave the truth
unknown." He knows that hiding the truth will
prolong the pestilence and ultimately destroy the
state. He discovers, however, that seeking the truth is
not without risk, but it is that very search that constitutes his greatness. He is after all Oedipus. It is he
who solved the riddle of the Sphinx, the answer being
man. In other words the search for truth and becoming human are, and have been since the days of
Oedipus, inextricably tied together. To be sure,
Oedipus' quest, in which his greatest obstacle is the
fear and ignorance of those about him, leads to great
pain and loss; the very thing that makes him great
leaves him vulnerable. He finds himself, as we do
with *Challenger* and Chernobyl, "mocked for the

very skills that prove me great." Oedipus could have ignored the imperative of the truth, could have given in to fear. We can too. But to do so is not to be fully human.

MEDICAL ETHICS IN THE AGE OF TECHNOLOGY

by

Tom L. Beauchamp

Tom L. Beauchamp

Tom L. Beauchamp is Professor and Senior Research Scholar at the Kennedy Institute of Ethics, Georgetown University.

Professor Beauchamp received a B.A. and an M.A. from Southern Methodist University, a B.D. from Yale University, and a Ph.D. in Philosophy from The Johns Hopkins University. He has been on the faculty of Georgetown University since 1970 and served as the Staff Philosopher of The National Commission for the Protection of Human Subjects of Biomedical and Behavioral Research of the National Institute for the Humanities from 1977 to 1979. He is the Series Editor (with Elizabeth Beardsley) of The Foundations of Philosophy *series for Prentice-Hall, the General Editor (with David Fate Norton and M. A. Stewart) of* The Critical Edition of the Works of David Hume *for Princeton University Press, and the Chairman of the Committee on Medicine and Philosophy of the American Philosophical Association.*

In addition to many articles in learned journals, Professor Beauchamp is the author of Philosophical Ethics, A History and Theory of Informed Consent *(with Ruth R. Faden)*, Medical Ethics: The Moral Responsibilities of Physicians *(with Laurence McCullough)*, Health and Human Values *(with Frank Harron and John Burnside), and* Principles of Biomedical Ethics *(with James F. Childress). He is coeditor of* Ethics in Public Policy, Contemporary Issues in Bioethics, *and* Ethical Issues in Death and Dying. *He has recently extended his investigation of ethical questions into the fields of business and journalism.*

MEDICAL ETHICS IN THE
AGE OF TECHNOLOGY

by

Tom L. Beauchamp

The primary objectives of the enterprise of bio-medicine have endured since antiquity: prolonging life, preventing pain, and preventing or alleviating illness, injury, and handicapping conditions. Despite this stability of purpose, in the last half century various scientific, technological, and social developments have caused rapid changes and altered moral perspectives. Simultaneously, public health and health policy have been in process of revolutionary change, in part caused by dramatic increases in health-care costs, new information about environmental and occupational hazards, and developing techniques in genetics.

I shall be arguing that these developments have challenged, and will continue for some while to challenge, traditional conceptions of the moral obligations of health professionals and of society. Often these problems are categorized as legal and political, but this conception is unduly narrow. We have had an unfortunate tendency in the United States to look to legal and regulatory approaches for our standards of conduct, but it has become progressively clear that the focus of statutory law, case law, and regulatory guideline is often misdirected. Many of the

problems I shall discuss, for example, are moral issues
about proper authority, communication, and personal
response, rather than abstract and disembodied issues
about proper legal standards of what must be disclosed and what constitutes malpractice. In short,
these are problems about the moral responsibilities
of physicians, not about their legal or political
responsibilities.

But moral theory also has its shortcomings. Despite
the distinguished history that medical ethics has
enjoyed—dating from the days of Hippocrates—
traditional codes and practices of biomedical ethics
have proved insufficient in the last two decades to
address numerous problems arising from modern
scientific research, biomedical technology, and
related social developments. We should not ignore
the historical record of traditional moral reflection
that has preceded us, but we also cannot lean on
empty sayings of the past as if they were the ground-
ing pillars of the present. My claim that historical
precedent is outdated needs justification, of course,
and I propose to begin with just such a study of historical traditions and their failure in the modern age.
After a brief look at the historical writings and
practices, I shall move to several topics of current
interest.

Medical Ethics from Hippocrates to the AMA

The history of the major writings of prominent
figures in ancient, medieval, and modern medicine
contains a rich storehouse of information about
cultural and theoretical commitments to the relation-

ship between professional and patient. But it is also a disappointing history from the perspective of contemporary biomedical ethics. It shows primarily how inadequately, and with what measure of hostility and insularity, problems of truthfulness, confidentiality, fairness, disclosure, and consent were framed and discussed prior to the twentieth century.

The Hippocratic Oath (or Oaths, as some prefer to say) developed from this environment into a public pledge to uphold professional responsibilities. The oath fails, however, to address what are now considered fundamental issues of responsibility, including the roles of communication, disclosure, and the giving of permission in the patient-physician relationship. Consent by patients, privacy, confidentiality, resource allocation, and the right to refuse treatment go unmentioned, and most topics and problems in modern medical ethics are ignored or given but passing notice.

The Corpus Hippocraticum bluntly advises physicians of the wisdom of "concealing most things from the patient, while you are attending to him . . . turning his attention away from what is being done to him; . . . revealing nothing of the patient's future or present condition." (*Decorum*, XVI, in W.H.S. Jones, trans., *Hippocrates*, Vol. 2, Harvard University Press, 1923-31, pp. 297, 299. See also Ludwig Edelstein, "The Hippocratic Oath: Text, Translation, and Interpretation," *Supplement to the Bulletin of the History of Medicine* 30, Supplement 1, The Johns Hopkins University Press, 1943.) The physician is portrayed in these writings as the one who commands and decides, while patients are conceived as compliant individuals who should place themselves in physicians' hands and obey commands. The Hippocratic tradition features the overriding importance of authority and shows little appreciation of patient needs apart from medicine's conception of the patient's medical needs. I shall refer to this structure of thinking about the responsibility of physicians as "Hippocratic beneficence"; that is, a conception of responsible medical behavior as resting entirely on acting for the patient's medical best interest.

Percival's Medical Ethics. The most celebrated work on medical ethics after Hippocrates was written by Thomas Percival (1740-1804). He published his landmark *Medical Ethics* in 1803—although it had first been drafted in 1794 for hospitals and medical charities. A diverse work on both ethics and etiquette, this volume is aligned with Hippocratic beneficence in its sections on "professional conduct" in encounters with patients. (*Medical Ethics; or a Code of Institutes and Precepts, Adapted to the Professional Conduct of Physicians and Surgeons*, S. Russell, 1803. The now more readily available edition is Chauncey D. Leake, ed., *Percival's Medical Ethics*, Robert E. Krieger Publishing Company, 1975.)

Like the Hippocratic physicians, Percival moved from the premise that the patient's best medical interest is the proper goal of the physician's actions to descriptions of the physician's proper deportment, including traits of character that maximize the patient's welfare. Recognizing the dependence of patients, physicians were counseled to discard feelings of pride and dignity, attending strictly to the patient's medical needs. Authority directs the physician to role responsibilities, dictated by the profession's understanding of its obligations, which are invariably focused on the patient's medical welfare. Again we see the triumph of the beneficence model as the framework for medical ethics.

The AMA and Its Legacy. Percival's work served as the model for the American Medical Association's (AMA) first Code of Medical Ethics in 1847. Many whole passages were taken verbatim from Percival, including the prescription of beneficent deception for patients who suffer from "gloomy prognostications." (See American Medical Association, *Proceedings of the National Medical Conventions, Held in New York, May 1846, and in Philadelphia, May 1847* [Adopted May 6, 1847, and submitted for publication in Philadelphia, 1847], p. 94.)

For the next century, and a few years beyond, the basic vision of medical beneficence founded by Hippocrates would remain intact as the dominant view of the physician's moral responsibility.

The Arrival of a New Medical Ethics

Shortly after the middle of the twentieth century, however, a major transformation began to take place: Forces in ethics and health policy having roots *external* to the professional concerns of physicians began to be influential. Sometimes they were greeted as unwanted alien intrusions; but in other cases they were greeted in medicine with open admiration. Of these influences, case law and other legal sources were most important. Generally written by lawyers, these reports functioned to alert health-care professionals to new legal developments and to potential malpractice risk. At about the same time that complex developments in case law were provoking reactions in medical ethics and the health-care community generally, events at Nuremberg, Helsinki, NIH, and FDA began cumulatively to have a revolutionary impact on research ethics.

In the 1960s problems of medical ethics were viewed in health-care circles as largely legal issues. New terms like "bioethics" were still not invented in the 1950s and early 1960s. During approximately this same period, there were parallel events of importance in medical technology, some with far-reaching implications for public policy. For example, dialysis emerged in Seattle in 1962 with widespread publicity surrounding moral problems in the selection of patients as beneficiaries of this technology. Later in the 1960s, this issue spread to kidney and heart transplantation, and was joined with an increasing focus on problems of access to health care. These spectacular biomedical and political developments retrospectively appear to have been necessary, although not sufficient, conditions of a new era of moral reflection on health care.

Many hypotheses could be invoked to explain why and how case law and bioethics came to address new, almost entirely unexplored subjects, especially having to do with the morality of patient care, and also why and how law and ethics influenced each other. Perhaps the most accurate explanation is that law and ethics, as well as medicine itself, were all affected by issues and concerns in the wider society about individual liberties and social equality, made dramatic by increasingly technological, powerful, and impersonal medical care.

Although this thesis would be difficult to sustain without situating it in the context of a careful and extensive socio-historical analysis of the period, it seems likely that increased legal interest in the right of self-determination and increased philosophical interest in the principle of respect for autonomy and individualism were but instances of the new rights orientation that various social movements of the last thirty years introduced into society. The issues raised by civil rights, women's rights, the consumer movement, and the rights of prisoners and of the mentally ill often included health-care components such as reproductive rights, rights of access to abortion and contraception, the right to health-care information, access to care, and rights to be protected against unwarranted human experimentation. These urgent societal concerns helped reinforce public acceptance of the notion of rights as applied to health care.

The rise of interest in bioethics in the second half of the twentieth century may have been as much a result of complex social forces changing the role and status of American medicine as a reaction to specific legal developments. An obvious example is found in the 1972 "Patient's Bill of Rights" of the American Hospital Association. (*Hospitals* 47 [February 1973], p. 41.) This bill was passed in large part because of consumer pressures for better care and facilities, as well as for

more appropriate standards of respect. The Patient's
Bill of Rights was connected to various consumer and
civil-rights movements that were everywhere demanding increased rights to make free and informed
decisions. The result of such developments has been
to introduce both confusion and constructive change
in American medicine in the struggle to meet un-
precedented challenges to traditional medical ethics.
At the same time, responsibility for fixing the terms of
responsible medical ethics has shifted from the physician's turf to that of the wider society. (On the
patient's rights movement generally, its origins and
influences, see Annas, "Patients' Rights Movements,"
Encyclopedia of Bioethics, Vol. 3, pp. 1202-5. See also
Bradford Gray, "Complexities of Informed Consent,"
*Annals of the American Academy of Political and So-
cial Science* 437 [1978], pp. 37, 40; Bernard Barber,
"Compassion in Medicine: Toward New Definitions
and New Institutions," *New England Journal of
Medicine* 295 [1976], p. 939; and Willard Gaylin,
"The Patient's Bill of Rights," *Saturday Review of the
Sciences* 1 [February 1973], p. 22.)

It is doubtful that the lines of influence between
medical ethics and the social movements of the 1960s
and 1970s can be easily untangled. However, this
much is clear: Once the flood was flowing in the
direction of a new medical ethics there seemed no
way to stop it. Issues arose from every quarter. A
massive tangle of problems with special literatures
emerged on the just allocation of medical resources,
informed consent, abortion, the definition and determination of death, euthanasia and the prolongation of
life, the use of behavior control techniques, and

genetic intervention and reproductive technologies. In the decade from 1962 to 1972, the old ideas of medical ethics began to crumble and the new emerged with vigor—so much so that terms like "bioethics," "moral problems in medicine," "biomedical ethics," and the like drowned the term "medical ethics," as if to signal the dawning of a new subject matter in a new era.

In one major respect the new "bioethics" is genuinely new. It is an interdisciplinary field uncontrolled by and heavily distanced from professional codes and practice. Contemporary medical ethics is the offspring of these recent historical developments, only incidentally descended from older forms of thinking. These recent concerns were never the focus of the great writings and teachings in clinical ethics, theology, or any discipline traditionally responsible for addressing the search for moral truths in health care.

Changing Models of the Patient/Professional Relationship

As long as what I have called the beneficence model remained the unchallenged model for medical ethics, physicians were able to rely almost exclusively on their own judgment about their patients' needs for medication, information, and consultation. Gradually, however, physicians were able to do more with new forms of technology than their patients might want them to do. Medicine was confronted with the need for a patient to make an independent judgment and consequently with values in those judgments that

were sometimes external to medical values. At the
same time, the external forces I have cited began to
challenge various assumptions in the beneficence
model of professional responsibility. These various
forces, working in tandem with the social movements
mentioned above, introduced what I shall call an autonomy model of the professional's responsibility; that
is, a patient-centered medical ethics that emphasized
autonomy rights rather than professional obligations
of care.

The central problem of authority in these discussions is whether an *autonomy model* of medical
practice that gives premier decisionmaking authority
to patients should be allowed to gain practical priority
over a *beneficence model* that gives authority *to
providers* to implement sound principles of health
care. The "autonomy model" here refers to the view
that the physician's responsibilities to the patient of
disclosure, confidentiality, privacy, and consent-
seeking are established primarily (perhaps exclusively) by the principle of respect for autonomy.
(The discussion of these models is derived from the
formulations of them in Tom L. Beauchamp and
Laurence McCullough, *Medical Ethics: The Moral
Responsibilities of Physicians*, Prentice-Hall, Inc.,
1984, esp. Chapter 2.)

The contrast is the following: The beneficence
model depicts the physician's responsibilities of disclosure and consent-seeking as established by the
principle of beneficence, in particular through the
idea that the physician's primary obligation (surpass-
ing obligations of respect for autonomy) is to provide
medical benefits. The management of information is

understood, on the latter model, in terms of the management of patients ("due care") generally. That is, the physician's primary obligation is to handle information so as to maximize the patient's medical benefits.

A typical handling of the conflict between these two models is found in the following statement by the recent President's Commission for the Study of Ethical Problems in Medicine and Biomedical and Behavioral Research.

"The primary goal of health care in general is to maximize each patient's well-being. However, merely acting in a patient's best interests without recognizing the individual as the pivotal decision-maker would fail to respect each person's interest in self-determination. . . . When the conflicts that arise between a competent patient's self-determination and his or her apparent well-being remain unresolved after adequate deliberation, a competent patient's self-determination is and usually should be given greater weight than other people's views on that individual's well-being. . . .

"Respect for the self-determination of competent patients is of special importance. . . . The patient [should have] the final authority to decide." (*Making Health Care Decisions*, U.S. Government Printing Office, 1982, Vol. 1, pp. 26-27, 44.)

The idea of a patient's giving an autonomous authorization or refusal suggests that a patient or subject does more than express agreement with, acquiesce in, yield to, or comply with an arrangement or a proposal. He or she actively authorizes and does not merely assent to a treatment plan through submission to a doctor's authoritative order. Thus, a patient would have to substantially understand the circumstances, decide in substantial absence of control by others, and intentionally authorize a professional to proceed with a medical or research intervention. This is certainly an arduous demand in medicine, and not one that many patients are well suited to perform, especially those who are seriously sick or suffering. (Ruth R. Faden and Tom L. Beauchamp, *A History and Theory of Informed Consent*, Oxford University Press, 1986, Chapters 7-8.)

Professionals and patients alike tend to see the authority for some decisions as properly the patient's and authority for other decisions as primarily the professional's. It is widely agreed, for example, that the decision whether to have elective surgery involving significant risk is properly the patient's but that decisions about whether and under what conditions to administer a sedative to a frightened patient screaming in an emergency room is properly the physician's. However, many cases are far more ambiguous as to proper decisionmaking authority—for instance, who should decide which aggressive therapy, if any, to administer to a cancer victim or whether to prolong the lives of severely handicapped newborns.

Judgments that the patient ought to serve as the primary decisionmaking authority are not always justified by a principle of respect for autonomy, because these judgments may rest on the principle that the patients' *welfare* is maximized by allowing them to be decisionmakers. The idea that authority should rest with the patients or subjects is justified, from this perspective, by arguments from beneficence to the effect that decisional autonomy by patients enables them to survive, heal, or otherwise improve their own health. These arguments range from the simple contention that making one's own decisions promotes one's psychological well-being to the more controversial observation that patients generally know themselves well enough to be the best judges, ultimately, of what is most beneficial for them.

The logic of the argument thus far is that we are poised at this moment in the history of medical ethics between an old and a new model of the physician's primary responsibility. Let me now turn to some of the implications of this situation for policy and practice.

The Quality and Necessity of Consent

First, problems about the quality and adequacy of consent probably cannot be resolved unless conventional disclosure rules are abandoned and a shift occurs to the quality of understanding present in a "consent." Dr. Jay Katz's recent book *The Silent World of Doctor and Patient* (The Free Press, 1984) is symbolic of the revolutionary potential in the demand for more patient autonomy and a weakening of controls exerted by medical professionals:

"I do not mean to suggest that physicians have not talked to their patients at all. Of course, they have conversed with patients about all kinds of matters, but they have not, except inadvertently, employed

words to invite patients' participation in sharing
the burden of making joint decisions." (Pp. 3-4,
28.)

Until recently, roughly the 1960s, the beneficence
model not only dwarfed any nascent autonomy model
in medical practice, but led to an environment in
which autonomy figured insignificantly or not at all
in reflections about disclosure. The lines of authority were roughly what they were in Percival's time—
the beneficence model being overwhelmingly pre-
dominant.

But there are signs of change. One area undergoing
rapid transformation is the legal concept of *the
therapeutic privilege*. In clinical settings this
privilege has long been used to justify not obtaining
consent and has elicited a particularly furious recent
exchange over whether autonomy rights can be validly overridden for paternalistic reasons. When framed
broadly, the therapeutic privilege can permit physicians to withhold information if disclosure would
cause *any* countertherapeutic deterioration, however
slight, in the physical, psychological, or emotional
condition of the patient. When framed narrowly, it
can permit the physician to withhold information if
and only if the patient's knowledge of the information
would have serious health-related consequences—for
example, by jeopardizing the success of the treatment
or harming the patient psychologically by critically
impairing relevant decisionmaking processes.

The narrowest formulation is that the therapeutic
privilege can be validly invoked only if the physician
has good reason to believe that disclosure would

render the patient incompetent to consent to or refuse the treatment; that is, would render the decision nonautonomous. To invoke the therapeutic privilege under such circumstances does not conflict with respect for autonomy, because an autonomous decision could not be made in any event. However, broader formulations of the privilege that requires only "medical contraindication" of some sort do operate at the expense of autonomy. These formulations may endanger autonomous choice altogether, as when the invocation of the privilege is based on the belief that an autonomous patient, if informed, would refuse an indicated therapy for what the medical professional views as incorrect or inappropriate reasons.

Such paternalism is sometimes resolutely and properly contested, because it threatens the basic values underlying the obligation to obtain the patient's permission. Loose standards for withholding information and for rationality can permit physicians to climb to safety over a straw bridge of speculation about the psychological consequences of information. In short, there is a significant potential for abuse of the therapeutic privilege because of its inconsistency with the patient's right to know and to decline treatment. (See Faden and Beauchamp, *A History and Theory of Informed Consent*, p. 38.)

The autonomy model has begun to rivet attention on the patient's *right* to make fundamental choices, and this shifts the burden of proof from the physician's obligations as beneficent caretaker to the physician's obligations as guardian of the patient's autonomy. This emphasis on rights expresses the real revolution that has occurred as a result of the rise of the autonomy model. However, the autonomy model is still a novel and provocative idea in medicine—and perhaps a model that must always compete with the beneficence model in the care of patients.

Refusal and Withholding of Treatment

The obligation to respect the autonomy of persons so that they are free and informed has proved particularly controversial when applied to the right to refuse *life-sustaining* medical interventions, where this obligation can easily be used to defeat normal presumptions in the medical community in favor of providing care and sustaining life. Modern medical advances have confronted us with this problem: Human bodies can now be kept alive for years without whole brain death and without consciousness. Dying can be extended indefinitely. Diseases that would have efficiently killed in the past now may paralyze or slowly sap life away. Here medicine may save us, but at an unacceptable cost to our quality of life. As a consequence, medicine has helped return us to certain classical debates in the Greco-Roman world over the acceptability of suicide, assisted suicide, and euthanasia. (I note that laws in Texas have never forbidden suicide and that not until 1973 did the law in this state forbid assisting in an act of suicide.)

On the one hand, recognition of a strong patients'-rights premise as authoritative in matters of refusing treatment could have a pervasive and unsettling effect on hospitals, where behaviors of foregoing treatment are still generally viewed as suspect and disruptive (despite the new views about patients' rights and respect for autonomy). On the other hand, we now live in an era in which state legislatures write natural death statutes, the Hemlock Society writes do-it-yourself guides for patients who wish to commit suicide, and voluntary euthanasia is increasingly viewed as an autonomy right.

A logical question to ask in this setting concerns the exact demands the principle of respect for autonomy makes in the decisionmaking context—for example, as to requirements that certain kinds of information be disclosed on the basis of which a patient might choose to refuse a recommended therapy. Another question concerns the restrictions society may rightfully place on choices by patients or subjects when these choices conflict with other values. For example, the patient might need scarce resources or might be abandoning small children in a foolish act of suicide. The courts have generally ruled that competent individuals have the right to refuse life-sustaining interventions, but exceptions have been made when third parties are involved and when full disclosure of negative information might lead a patient to irrationally refuse essential therapies. Some courts have also attempted to protect the "autonomy" of incompetent individuals by the doctrine of substituted judgment, which asks what an individual would want were he or she competent. This doctrine too has raised a hornet's nest of legal, moral, and conceptual problems.

Almost certainly the major problem to arise in recent years, and one that may substantially change traditional understandings of the patient/provider relationship, is the withholding or withdrawal of

hydration or nutrition. Fairly early in the recent
discussion of withholding or withdrawing high-
technology medical equipment, in the mid- to late-
1970s, it was common to hear both physicians and
lawyers say that respirators and dialyzers can be
removed, but not intravenous fluids (IVs). The
thought seemed to be that respirators and dialyzers
were extraordinary and IVs ordinary. But when the
distinction between the extraordinary and the ordinary was thoroughly undermined by careful conceptual analysis of these notions, this undercut the
rationale for the insistence on the use of IVs while
permitting discontinuance of the respirator. Around
1984 the question of withholding fluids and nutrition
became a troublesome practical ethical dilemma, one
suggesting that we more carefully examine our
criteria of what counts as appropriate and necessary
medical care, especially when a treatment might be
burdensome to the patient, could fail to provide any
benefit, or could be unreasonably expensive in light
of the expected benefits.

Being able to supply artificial nutrition and hydration in the form of properly balanced liquid diets
through means such as flexible plastic tubing is itself a
technological advance made in the past forty years.
Intravenous tubes, nasogastric tubes, and surgically
implanted gastrostomy tubes are technological
marvels that some see as typical forms of medical
treatment. They see these treatments as relevantly
resembling respirators for breathing, for example.
But where they see a relevant resemblance, others
see a sharp difference. Many regard water and food as
basic forms of health care that can never be justifiably

withheld or withdrawn in the way respirators and many optional forms of treatment are subject to discretionary judgments. The issue is muddied by the fact that we are often talking about high-tech medications such as enteral alimentation or parenteral hyperalimentation—and here there may be a debate about whether they should be used at all.

An act of withholding or withdrawing hydration and nutrition is ordinarily meant to end life—not, presumably, as an act of killing, but rather as an act of allowing to die. Those who see this act as hastening the time of an imminent death by not using a half-way technology that cannot reverse damage already done have no difficulty with these practices of withholding. But others see such acts as the proximate cause of death, thus raising questions about the appropriateness of killing, of euthanasia practices, and of the involvement of health-care professionals in such activities. (For these arguments, see John J. Paris, "When Burdens of Feeding Outweigh Benefits," *Hastings Center Report* [February 1986], pp. 30-32; "The Six-Million-Dollar Woman," *Connecticut Medicine* 45 [1981], pp. 720-721; and the articles referenced below by Callahan, Derr, Siegler, and Weisbard.)

There is presently no consensus resolution of this problem, although there are numerous court decisions and pronouncements by professional organizations. One prominent view was advanced by the aforementioned President's Commission, which issued a large volume on *Deciding to Forego Life-Sustaining Treatment* in March, 1983. The Commission maintained that in the medical setting today, "for almost any life-threatening condition, some invention can now delay the moment of death," but found that no particular treatments, even such "ordinary" ones as special feeding procedures, were mandatory in all cases. (U.S. Government Printing Office, 1983, pp. 1, 90.)

For example, in the instance of permanently unconscious patients the Commission held that the only conceivable benefit is sustaining the body, and this is no real benefit at all. Not only can costs to family and society be burdensome but, in the Commission's view, they can violate the patient's autonomy if one knows what the patient formerly expressed about what should be done were he or she ever placed in such a difficult position. (*Ibid.*, p. 190.) The Commission even suggests that "only rarely should a dying patient be fed by tube or intravenously." (*Ibid.*, p. 288.)

The American Medical Association has also made two influential pronouncements on this general subject. In 1982, the Association's Judicial Council (its ethics arm) stated as follows: "Where a terminally ill patient's coma is beyond doubt irreversible and there are adequate safeguards to confirm the accuracy of the diagnosis, all means of life support may be discontinued." (*Current Opinions*, pp. 9-10.) On March 15, 1986, in "Opinion," the AMA's Council on Ethical and Judicial Affairs clarified the 1982 statement, making specific reference to the problem of not providing nutrition and hydration. The Council held that "Life-prolonging medical treatment includes medication and artificially or technologically supplied respiration, nutrition, or hydration." Although the AMA's Council has made its pronouncement in an extremely narrow range—confirmed irreversible coma—it is unambiguous that nutrition and hydration may be discontinued.

A number of medical societies have made pronouncements on this subject that pertain to autonomy rights, making a direct appeal to the priority of the autonomy model over the beneficence model. For example, the Massachusetts Medical Society passed a resolution on July 17, 1985, that reads as follows:

"The Massachusetts Medical Society recognizes the autonomy rights of the terminally ill and/or vegetative individuals who have previously expressed their wishes to refuse treatment including the use of intravenous fluids and gastrointestinal feeding by tube and that the implementation of these wishes by a physician does not in itself constitute unethical medical behavior provided the appropriate medical and family consultation is obtained."

Such pronouncements by professional organizations are usually confined to permanently unconscious patients. But many commentators have gone much further, extending the discussion to the withdrawing of fluids from the terminally ill and even from some nonterminally but seriously ill patients. (See D. W. Meyers, "Legal Aspects of Withdrawing Nourishment from an Incurably Ill Patient," *Archives of Internal Medicine* 145 [Jan. 1985], pp. 125-128.)

Clearly our intuitions and considered judgments are divided on these matters. When Karen Quinlan's

father determined to remove her from a respirator, he
was asked if he wished to remove her intravenous
feeding as well. He refused, on grounds that the feeding was her nourishment. Even if there is no clear
morally relevant difference between respirators and
nutritional supports, we have difficulty with *saying*
that there is no relevant difference and living with
the consequences. Many, especially medical professionals, find it difficult to starve someone to death.
The symbolic and emotional power of the fundamental idea of feeding the sick and the dying is among
the most poignant dimensions of our social system.
(See Daniel Callahan, "On Feeding the Dying," *The
Hastings Center Report* [Oct. 1983], p. 22.)

Some claim that more than emotion is at work
here: They claim that there *is* a morally relevant
difference between denying food and fluids and denying medical therapies. For one thing, there is a
finality and certainty about causing death when nutritional needs go unmet. They also see the intrusion of
nutrition-denial into medical institutions as an attack
on the image of health-care providers and on their
professionalism. They believe the medical profession
should not be in the business of balancing quality-of-
life considerations and cost concerns against the
compassionate standards of medical care. They see
such judgments as value laden and ad hoc, running
against the grain of the physician both as scientist and
as one who professes a firm set of moral values of
justice, care, and commitment. Some quite specific
worries accompany these general worries; e.g., the
idea is troublesome that physicians, families, courts,
or various other third parties can rightly make a deci-

sion that the "burdens" of continuing fluids and nutrition can outweigh the "benefits" of sustaining life. (See Mark Siegler and Alan J. Weisbard, "Against the Emerging Stream: Should Fluids and Nutritional Support Be Discontinued?" *Archives of Internal Medicine* 145 [Jan. 1985], pp. 129-31; and Patrick G. Derr, "Why Food and Fluids Can Never Be Denied," *The Hastings Center Report* 16 [Feb. 1986], pp. 28-30.)

In considering a variety of treatment denials, the courts have exhibited considerable confusion about how to handle these cases. Appellate courts have reversed lower courts right and left. There is no clear and applicable law, and so we should not be surprised that the bench reflects our own ambivalence and uncertainty. (See, for example, *Bouvia v. County of Riverside*, No. 159780 [Super. Ct., County of Riverside, Calif., Dec. 16, 1983], and *Bouvia v. Superior Court, Los Angeles County*, 2nd Civ. No. B019134 [Cal. Ct. App. April 16, 1986]; *In the Matter of Mary Hier*, 18 Mass. App. Ct. 200 [Essex Div., Probate Ct., 1984]; *Corbett v. D'Alessandro*, No. 85-1052 [Fla. App. 2nd Dis., April 18, 1986]; *In re Nancy Ellen Jobes*, No. C-4971-85E [Super. Ct. N.J., April 23, 1986]; *In re Conroy*, 98 N.J. 321, 486 A.2d 1209 [N.J. 1985]; and *Brophy v. New England Sinai Hospital*, Docket No. 85E0009-G1, slip. op. [Mass. Probate and Family Ct., Norfolk, Oct 21, 1985].)

The Bouvia case, cited above, is an excellent example. Ms. Bouvia is severely handicapped and suffering from cerebral palsy. She sought a permanent injunction against a hospital that was feeding her against her wishes. A lower court concluded that a restraining order on the hospital would have "a devastating effect" on other patients in the institution and on other handicapped persons. This court concluded that "She does have the right to terminate her existence but not while she is nonterminal with the assistance of society." The Appellate Court in this case held that "the trial court seriously erred by basing its decision on the 'motives' behind Elizabeth Bouvia's decision to exercise her rights" and went on to a strong defense of her rights, especially of her request to have a nasogastric tube removed. In a separate concurring opinion, Justice Compton argued that "she has an absolute right to effectuate [her] decision. This state and the medical profession instead of frustrating her desire, should be attempting to relieve her suffering by permitting and in fact assisting her to die with ease and dignity. The fact that she is forced to suffer the ordeal of self-starvation to achieve her objective is in itself inhumane." (*Bouvia v. Superior Court, Los Angeles County*, 22-27, and Concurring Opinion, 2.)

Meanwhile, physicians and health-care administrators are disquieted by the prospect of withdrawing hydration and nourishment, and state laws often conflict with the pronouncements of medical professional societies. Increased litigation is a certainty. Not everyone is ambivalent and uncertain about these matters, but we do seem confused as a society.

The Idea of a Right to Health Care

The 1965 law that established Medicare for the aged and the extension of coverage to dialysis patients in 1972 were revolutionary developments in

American society. These mechanisms for funding the delivery of health care were historically as well as symbolically important. They have come to stand for both substantive change in our traditional use of free-market means of financing health care and difficulties in achieving workable and tidy programs of access and cost containment.

These issues had emerged before we were well prepared for them. The problem was made even more difficult to confront because of a new language of *entitlement* that accompanied the language of patients' rights and autonomy. In previous eras health care for the needy was guided by the virtues of charity, benevolence, and compassion, and the rhetoric used was therefore not the language of rights and justice to which we have become so accustomed today. The underlying problem is that in our technologically advanced societies venerable ways of handling health-care needs through appeals to virtue have faltered in the face of dramatically increased costs of and needs for care. In the old days, various community, charitable, and religious institutions handled our comparatively meager needs for health care. But the health needs of a significant sector of the population went unmet as communities simultaneously grew more impersonal and biomedicine exploded with new developments and needs for funding.

Subsequently, many have become concerned about the possibility that the language of a *right* to health care will entail uncontrollable expenditures from the public purse. They worry that our early funding of end-stage renal disease may turn out to be the thin edge of a wedge leading to unjustifiably expensive

federal involvement in the health-care sector. Others
debate the fairness of further shifts in tax burdens on
corporate and individual incomes in order to provide
assistance. Many are troubled by the possibility that a
theory of political equality might be used as the basis
of a theory of economic equality of distribution. Discussions today of "equal access" and "the right to
health care" are the direct descendants of advances in
the technology of health care, and we are still in
search of a consensus position and rationale for our
policies.

Perhaps the policy at the present time most in need
of assessment is the DRG (Diagnosis Related Groups)
system of payment for Medicare patients. It is a new
phenomenon in medicine when the "cost overruns"
of a single member of a staff might undermine an
entire program in a hospital and when moral problems arise regarding whether patients receive too
many resources or too few resources based not on a
physician's judgment but rather on a system of
reimbursement by prior set amounts. This system,
together with cost containment and reduction
programs, has immediate impact on the practice of
medicine—not only regarding how much we pay for it
but how much we value it and how many risks we are
willing to take in practicing it.

For example, as a result of the DRG system, conceptions of ideal or even proper care may have to
be compromised to accommodate the demands of
utilization review committees and expected reimbursement. Some will maintain that ideal care
need not be sacrificed, but it is unrealistic to suppose
that physicians will be allowed to exceed prospective

payment scales with frequency or without well-reasoned justification. Again many hospitals will surely be tempted to eliminate the most expensive patients in any given category.

To ask physicians to be cost-aware and to aim at a particular level of care is to ask them to depart from a central platform in traditional medical ethics: the primary commitment to the patient. Even if clinicians are not the primary decisionmakers regarding acceptable expenditures, someone must play that role, and physicians will not be exempt from the effects of these decisions. Labels such as "policy choices" and "cost containment" may be invoked to dignify these decisions, but they straightforwardly involve *moral* choices: They are assessments of what is in the overall public good.

That heart transplants cost too much and therefore do not warrant funding is similarly a moral, and not merely an economic judgment. The artificial-heart program presents some of the most interesting value choices around today. First, it shows how clearly a medical "benefit" is seen to be a value-laden idea in this example. Is the artificial heart a beneficial therapy? Was Barney Clark best served by the concentrated and caring treatments that he received? Secondly, is the sponsoring of an artificial-heart program the best use of available resources? When no such resources existed, there were no such choices to be made. But now we encounter them at every turn. (See Samuel Gorovitz, "The Artificial Heart: Questions to Ask, and Not to Ask," *The Hastings Center Report* [Oct. 1984], pp. 15-17.)

These general reflections on contemporary funding

problems can be illustrated by a concrete case, one
pending at the present time in the courts. This case is,
I believe, a typical case of new moral problems
presented by the DRG system. On June 7, 1985, the
Inspector General of the Department of Health and
Human Services issued an audit report reviewing
Medicare payments for assistant-surgeon services during cataract surgery. The Inspector General concluded that the Medicare program could realize
savings of about $150 million to $200 million over a
five-year period if there were "a national policy to exclude Medicare coverage of assistant-surgeon charges
on routine cataract operations." (Richard P. Kusserow,
Office of the Inspector General [DHHS], "Memorandum to Carolyn K. Davis, Administrator, Health
Care Financing Administration" [June 7, 1985], p. 1.)

The current medical practice endorsed by most
cataract surgeons requires the services of an assistant
surgeon during cataract surgery (although some
surgeons use only surgical technicians). The Inspector
General concluded—based on consultation with
various hospital, medical, and administrative
personnel and data supplied by ophthalmic
surgeons—that an assistant surgeon is not medically
necessary during routine cataract surgery, and therefore recommended that the Health Care Financing
Association (HCFA) exclude the services of an assistant surgeon from Medicare coverage. The Inspector
General recommended allowing such use of an assistant only when it is demonstrably essential for a
successful operation and prior approval has been
obtained. He also held that DRG coverage already in-

cluded reimbursement for surgical technicians, and he found this coverage sufficient to satisfy the "reasonable and necessary" requirement of Section 1862(a) of the Social Security Act.

Eventually assistant-surgeons' fees for cataract surgery were disallowed and directives were sent from the Office of the Secretary of Health and Human Services to Medicare carriers denying fees to assisting surgeons. The final decision had a somewhat unusual wrinkle: Not only would the government not reimburse for assistants, it also would not pay for the primary surgeon *if* patients paid out of their own pockets for the assistants. The logic of this position is paternalistic, or at least strikingly protective: DHHS wishes to curb all "excessive" medical payments by patients in order to protect the patient from manipulation. That is, DHHS is striving to discourage excessive charges by physicians if the government is involved in *any form* of payment in the case, even if the patient rather than the government is paying the "excessive fee." Not surprisingly, the New York State Ophthalmological Society and the California Association of Ophthalmology have sued the Secretary of DHHS in this case.

The justifiability of this paternalistic maneuver by federal officials is not an issue I wish to address. I want only to point out how the desire to limit and deny fees produced by the DRG system raises profound moral problems of the sort we did not envisage prior to this system. For example, in the judgment of many cataract surgeons the absence of an assistant adds risk of harm to the patient (e.g., risk of damage to vital ocular tissues such as the corneal endothelium, the iris, and the posterior capsule of the lens) that is inconsistent with good medical practice. The new DHHS requirement forces one either to practice outside the reimbursement system or to accept a form of practice that one believes to contain unjustifiable risks to one's patients.

The effect of this ruling also seems likely to increase patient apprehension. Any physician conforming to good practices of disclosure to the patient would inform the patient of the increased risk. This in turn will make the patient more apprehensive and will present dilemmas about payment for the best possible services (by disenrolling from part B of Medicare in order to avoid the statutory prohibition) and possibly even about foregoing needed surgery. Physicians making such an explanation will believe, as will patients, that less than optimal care is being provided, and thus that there has been a decline in the quality of patient care. In a few cases physicians will believe that unnecessary complications in surgery, poor vision, blindness, and even death are causal results of the failure to provide an assistant.

The patient's powerlessness to hire an assistant also places the physician in the awkward position of not knowing whether there is an ethical obligation to *avoid* performance of the surgery without an assistant. This will occur if physicians strongly believe that an assistant is needed to reduce risk but the patient agrees to undergo surgery at the increased risk (for, say, financial reasons). Some physicians may believe that the only appropriate response is to refuse to perform the surgery, thus leaving the patient to find another surgeon. Here we have a moral dilemma, with no clearly preferable course of action.

There is a respect in which the DRG system merely forces into the open what has been present beneath the surface all along; namely, that we have conflicting goals that cannot be reconciled in our health-care system. For example, we profess to provide the best possible care for everyone and at the same time we seek to put in place programs of cost containment. Our expectations for quality care are infinite, our commitment to the resources needed to meet those expectations quite finite. The two commitments are fundamentally incompatible. These two goals express our ambivalence about what we desire, and part of what we are doing in reflecting on the acceptability of our present funding for health care is to tangle with our ambivalence over our precise moral responsibilities for contributing to the health-care system generally.

I conclude not that the government is wrong in pressing its case or that California and New York ophthalmologists are wrong in pressing their case against the government. My argument has been that changes in the system of funding are forcing profound dilemmas about standards of appropriate care, and thus changes in our views about medical ethics.

Developments in Genetics

I turn, finally, to an area of profound change in modern biomedical ethics: the extraordinary developments in modern genetics. I shall mention only a few of many prominent problems.

Molecular Genetics and Gene Therapy. Research in
molecular genetics and parallel developments in
diagnosis have been extraordinary in recent years,
and there likely will be even more spectacular results
in upcoming years. Several laboratories now pursue
strategies for correcting inborn errors of metabolism
using gene therapy, which involves the introduction
of a normal functioning gene into a cell where a defec-
tive gene is active. For a number of hereditary dis-
orders gene therapy is imminent—e.g. for Lesch-
Nyhan disease, which produces mental retardation
and uncontrollable self-mutilation behaviors, and
ADA deficiency (adenosine deaminase), which causes
severe combined immune deficiency in children. The
boy in the bubble in Houston was a well-known example of the latter.

Although certainly extraordinary as an approach to
human genetic disease, the surgery involved in gene
therapy will not depart in many respects from conventional medical practice and therapy. For example,
bone-marrow transplantation will be supplemented
by the addition of a properly functioning gene to as
many target cells as possible. This is gene addition
and compensation rather than gene repair. The objective is to compensate for nonfunctioning or malfunc-
tioning genes. These are also somatic treatments—
i.e., ones that treat an individual victim suffering
from the disease but without affecting offspring who
might suffer. In this respect the surgery is like any
surgery to correct a problem in the body—e.g.
removing organs, where the surgery will not affect an
offspring's organs.

Critics of these developments fear that if we once

begin to modify genes to remove congenital defects, we cannot stop short of engineering the species. For example, it is feared that there will be alterations in the human germ line, and thus that intentional genetic changes will be transmitted to future generations. However, at the present time only somatic cells are used, and no researchers plan to introduce alterations into the human germ line (or reproductive cells). Moreover, no one at present knows how to develop a reliable technology of germ-line modification; only very simple single-gene defects are today's candidates for gene therapy. There is no reason to expect that the addition of these genes will be passed on through the reproductive process, and it seems morally irresponsible to oppose responsible research and development on grounds that someday there might be inappropriate uses of the technology. On this argument, no technology of any sort could be produced, on grounds that it might be misused for evil purposes.

An ethical issue of obvious importance is how to weigh the risks and benefits of this new form of experimentation and treatment. Risk assessment is complicated by the uncertainty that clouds the likely outcome of gene therapy. There are virtually no laboratory animal models for the single-gene defects most likely to be involved early in this therapy; and it thus seems unlikely that therapeutic benefits will be convincingly demonstrated in the laboratory or will be well understood before attempts with humans. Once again, the best experimental animal from the point of view of research is the human animal. The best candidates for gene therapy will be those patients for whom there is no other effective or

promising therapy at the present time. Phenylketo-
nuria (PKU), for example, does not seem to be a likely
starting point because there are alternative forms of
treatment, however unsatisfactory. But these are involved matters that invite careful technical as well as
moral assessment.

A second issue concerns the selection of patients.
As with risk assessment, this issue is not novel; there
is a history of discussion pertaining to the selection of
subjects for scarce and expensive medical treatments
and for organ transplantation. In the early stages of
gene therapy it is not likely that there will be a large
number of candidates, but the patients will grow in
numbers and the issue will gradually grow in importance. My colleague LeRoy Walters tells me that
he knows of one family that has moved to San Diego
where they believe the research being done holds out
the greatest prospects for helping their young son.
Presumably they are first in line, but is being first in
line the right method for selecting experimental
patients? Justice in the selection of these subjects will
be difficult to achieve, just as it has been elsewhere.

A third issue is that old menace mentioned previously, informed consent. Candidates for gene therapy
will need to have an unusually good understanding of
the nature of that therapy, of bone-marrow transplantation, of how risk assessments are done, and of how
much uncertainty is involved. Since the territory is
principally virgin, this information will be difficult to
package and present so that the patient can
comprehend the nature of the choices to be made.
But a sound understanding of the promise, the

problems, and the unknowns seems essential—no less so for the patients or their parents than for the physicians involved. (I am heavily indebted for the observations in this section to my colleague Dr. LeRoy Walters. See his "The Ethics of Human Gene Therapy," *Nature* 320 [20 March 1986], pp. 225-227. Another useful presentation of many of these issues is found in the President's Commission for the Study of Ethical Problems in Medicine and Biomedical and Behavioral Research, *Splicing Life* and *Screening and Counseling for Genetic Conditions*, Government Printing Office, 1982 and 1983.)

The Art of Medical Genetics. Also of interest as an outgrowth of genetics is the approach taken by physicians in counseling patients and parents. Unlike the novelty involved in gene therapy, here we already have a body of practitioners and a practice or set of practices that can be critically examined. The problem is that there is substantial disagreement even among participating physicians as to how to handle patients and parents. Consider a few examples.

First, how often does one fully disclose to patients or parents, rather than partially obscuring information or not disclosing at all? Do you disclose the fact of XY genotype in a female? Do you explain to a patient about controversial or ambiguous laboratory results, or about conflicting diagnostic findings? If one spouse is likely to blame or punish the other spouse for being a carrier, when both could be carriers, do you come forward with information about who the carrier is? There is little consensus at the present time regarding such matters as whether to disclose to patients against their wishes or about various forms of disclosure to relatives.

Huntington's disease—a genetic disease which attacks adults and often results in dementia—is a typical case where a patient can be given a plausible diagnosis, but without any prevention or therapy. Substantial questions surround whether the autonomy model should override use of a beneficence model and clinical judgment regarding the patient's ability to confront and handle results of the genetic tests, which can be quite devastating. Consider a clinician managing a patient who wants the test in order to commit suicide if the results are positive. Whether a physician has a duty to honor a patient's request for this reason turns on whether it is believed that the patient is competent to make such a decision as well as on the physician's moral views regarding the taking of human life and the role of medicine as facilitating the process. All are complicated moral issues without tidy resolutions.

Secondly, and relatedly, to what extent are you obligated to maintain confidentiality in the face of some form of obligation to a third party? This problem is most acute when one feels obligated to disclose, against a patient's express wishes, a diagnosis of Huntington's disease or of Hemophilia A to a fiancé or to relatives at risk. Again, there is at present only controversy and no consensus about how to handle such matters. There are also the staple cases of whether there should be a disclosure to a husband in a false paternity case, thus overriding a mother's right to confidentiality. Although medical geneticists generally exhibit strong lines of support for protecting the patient's confidentiality, these issues are not easily amenable to a blanket solution.

Throughout the history of Western medical ethics the rule has generally been observed that a physician's primary obligation is to benefit the patient and not to serve the next of kin, a legal guardian, or the state. Similarly, courts have traditionally excluded any role for the rights and interests of third parties in decisions concerning incompetent patients. We have generally applauded such rules on grounds of the protections they erect for vulnerable patients. However, the legal and medical institutions that have promoted and acted on such rules have not faced up to the broader social context in which the rights and interests of third parties inevitably play a role in judgments about the allocation of medical resources, about whether advance directives by patients will be recognized, about whether to withhold or withdraw life-sustaining treatments for incompetents. There are also moral duties to public-health authorities and police authorities.

Many third parties have both rights and responsibilities that simply cannot be ignored. They may be serving as the patient's fiduciary no less than the physician, and some of these parties may be significantly harmed if a physician or medical institution acts solely in the "best interests" of the patient. The family of a seriously ill or comatose person may, for example, suffer overwhelming emotional, psychological, and financial consequences if a family member is aggressively treated. What we need at the present time is a better conception of the role these third parties should and should not play in the process of medical decisionmaking—in particular, as surrogate decisionmakers or as parties whose interests are as deeply affected by the way the physician handles information as are the patient's interests.

Thirdly, many issues center around prenatal diagnosis. The problem of abortion, with its multiple dilemmas, is involved in various issues of how to counsel couples about some genetic problems, but there are other issues as well. For example, some parents refuse even to consider the possibility of an abortion before or after prenatal diagnosis. Should prenatal diagnosis be made available for these parents? Other persons, by contrast, use prenatal diagnosis purely for purposes of sex selection (unrelated to X-linked disease). These issues involve not only value choices about the purpose and proper use of prenatal diagnosis, but the proper roles for directive counseling and for the values of the physician as a guide for the value choices of patients. The issue of how to use prenatal diagnosis is among the most vexed of all issues of ethics in the community of physicians. Both in the United States and internationally, there is continuing controversy and no consensus about prenatal diagnosis for sex selection. (I am indebted to John Fletcher for a number of important observations and for showing me a preliminary report of an empirical study of these attitudes and beliefs. I am also indebted to Ruth Faden for insights into the importance of some of these issues in contemporary genetics.)

Finally, many questions surround genetic screening. New presymptomatic tests that identify a person as susceptible to a genetic disease (not as *having* or as with certainty later developing)—e.g. for Huntington's—are now being used clinically as forms of therapeutic experimentation, and a presymptomatic test for cystic fibrosis is under development. Extensive work is also being done on genetic tests for major chronic diseases, including malignant melanoma, various types of heart disease, breast cancer, Alzheimer's disease, and manic depression.

These methods of diagnosis and prevention promise enormous diagnostic and therapeutic benefits. But are methods of genetic screening for common diseases acceptable, and, if so, under what conditions? Must all screening be voluntary, or is some so important that we should make it mandatory? Are methods of genetic screening in the work place and for common diseases acceptable, and if so under what conditions? Again the role of third parties can be significant in answering such questions. Do government health departments or third parties such as employers or life insurers have any rights to genetic information against a patient's wishes? Should the information be routinely made available to the patient's physician, or should there be no access without the patient's consent? If a spouse or a relative is at risk for Huntington's disease, should he or she have access to any information without the patient's consent?

Screening in the Work Place. Consider in particular the use of screening in the work place. Present scientific capabilities for screening are likely to make many screening tests available in upcoming years. Genetic monitoring of workers for evidence of genetic damage

from work-place exposures is now administered by
testing employees for alterations in the genetic
material of body cells that are caused by substances in
the work environment. Such monitoring searches for
genetic damage—usually occupationally induced conditions that predispose individuals to further occupational disease, especially in circumstances where
there is compound exposure to multiple substances.
In principle, of course, such screening can also look
for inherited genetic traits not caused by occupational
exposures. Workers may have an increased risk of
genetic damage by virtue of inherited genetic traits.

The possibilities promised for this kind of screening
raise a nest of moral problems. Some of these issues
pertain to the reliability and predictive accuracy of
the tests, especially since they are used in an environment where there may be numerous possible causal
sources of the diseases thought to have been
detected. Poor testing may also lead to false catego-
ries and improper labels being given to those who
have been tested. But even if these more technical
problems can be overcome, there will still be an
ample set of moral problems to be confronted. For example, the labels can function in a discriminatory
fashion, especially when genetic predisposition follows ethnic lines.

Correlatively, there are substantial questions about
an employer's obligation to hire, promote, transfer,
and protect the worker. But the employer will also
have to ask and will want answers to many other
questions, such as whether workers may be fired,
transferred, or held fully responsible for ill outcomes
at the employer's discretion if they have the "wrong"

genetic predisposition. The employment-at-will doctrine is already being challenged in the courts, and it is likely that genetic information in the hands of employers and insurance companies will lead to intense public scrutiny of this venerable doctrine. Moreover, employers may also potentially be held responsible for erecting special protections in the work place for those who are hypersusceptible. From the employer's perspective, this is a nightmarish possibility—comparable to outfitting a work place with protections for even the genetically weakest members of the species.

Conclusion. Some scientists and funding sources now believe that perhaps within the next five years—and certainly before the turn of the century—we will have mapped the human genome (the characteristic set of genes of the human species). The enormously powerful new tools on the horizon—gene therapy, using genetic markers, new forms of screening, and mapping the genome—will inevitably pull physicians and biomedical researchers into public-policy issues no less than problems of medical ethics. One can hide from many of these issues now, because they are still relatively hidden issues. But if the entire human genome is soon to be mapped and we are to witness vast new diagnostic powers in medicine, then, like it or not, we will witness some form of what we can only hope will be a *brave* new world.

PROFESSIONAL AND BUSINESS ETHICAL AND MORAL VALUES IN THE AGE OF TECHNOLOGY

by

Jesse P. Luton, Jr.

Jesse P. Luton Jr.

Jesse P. Luton, Jr., is an attorney in Houston, Texas, where he is a partner in the law firm of Scott, Douglass & Luton. Educated at the University of Texas, he received his LL.B. with honors in 1948 following service in World War II as a Captain in the U.S. Army Corps of Engineers. After serving as a briefing attorney to the Supreme Court of Texas, he was an Assistant Attorney General of Texas. In that position, Mr. Luton participated in the representation of the State of Texas in the famous tidelands litigation and in other important cases in the Supreme Court of the United States and in other federal and state courts. For the next thirty years, he was an attorney in the law department of Gulf Oil Corporation, in which he held various positions, including General Counsel, until his retirement at the end of 1983 to join his present firm.

Mr. Luton is a member of the State Bar of Texas, in which he was Chairman of the Natural Resources Law Section and the Corporate Counsel Section. He is also a member of the American Bar Association, having been Chairman of the Natural Resources Law Section and a member of the House of Delegates. He has been active in matters relating to the University of Texas School of Law and was a member of the first Board of Visitors of that school.

Mr. Luton has been a speaker at various legal institutes on topics concerning litigation, corporate law department management, liability of corporate officers and employees, and energy law. He is the author of a number of published articles dealing with subjects in these and other areas.

PROFESSIONAL AND BUSINESS ETHICAL OR MORAL VALUES IN THE AGE OF TECHNOLOGY

by

Jesse P. Luton, Jr.

The Age of Technology has been a period in which more technological advancement has probably been made than in all of the prior history of the industrial world. Yet technology still is in its infancy, and the advancements and changes over the next few decades promise to exceed what we have already seen.

These technological developments have resulted in benefits which have significantly improved the quality of living and the ease and speed with which difficult and complicated tasks can be handled. However, they also have been accompanied by problems resulting from decreasing jobs and increasing unemployment, changes in the work force by displacement both of men by women and of unskilled workers, a growing concentration of business and a disappearance of many small businesses, stiffer competition in the United States and in foreign countries, and the move of United States business from a heavy industrial complex to a more technological and service-oriented complex.

In view of these and other changes which have been, are, and will continue taking place in the United States as a result in large part of the effects on

our society of technological advancements, the
subject of Traditional Moral Values in the Age of
Technology for this lecture series is timely and
appropriate.

The above problems accompanying, and in many
respects caused by, technology present questions as
to the responsibilities of management in dealing with
these problems and with employees affected by them.
These and other questions arising out of advanced
technology and the social consequences caused by it
involve ethical considerations. For example, what is
the ethical responsibility of management to provide
for the retraining or placement of employees who
are no longer needed because of changes or improvements in technology? Also, there are ethical and
moral considerations involved in the use of, or failure
to use, equipment that results in pollution of the
environment; in the use of animals and humans for experimental purposes; in the use of artificial birth
methods, surrogate mothers, and the control of the
sex of a baby; and in many other situations arising
from advances in technology and science. Each of
these matters is one that can be, and has been, a
subject for consideration and study from an ethical
and moral standpoint.

A recent illustration of this is a set of ethical guidelines offered by the American Fertility Society for
practitioners of new artificial birth methods. That
society consists of some ten thousand doctors and
scientists who work in reproductive technology. The
guidelines cover ninety-four double-columned pages
and involved eighteen months of work by an eleven-

member committee of doctors, scientists, lawyers, ethicists, and other experts. (*Wall Street Journal*, Sept. 9, 1986, p. 14.)

In this lecture, I will deal with recent challenges to time-honored ethical and moral values in business and professional life. Ethical or moral values are principles or standards "by which human actions are judged right or wrong." (*1986 Britannica Book of the Year*, p. 34.) Ethics "refers to imperatives regarding the welfare of others that are recognized as binding upon a person's conduct in some more immediate and binding sense than *law* and in some more general and impersonal sense than *morals*." (G. Hazard, *Ethics in the Practice of Law*, Yale University Press, 1978, p. 1.) "A moral value is one that is in harmony with what it means to be a human being in all that it implies." (E. Stevens, *Business Ethics*, Paulist Press, 1979, p. 15.) An "ethical businessman" has been defined "as a man who is the full measure of what a man should be." (P. Heyne, *Private Keepers of the Public Interest*, McGraw-Hill, 1968, p. 113.)

The ethical and moral values of those in business and the professions were the subject of a 1985 survey by the Roper organization. One thousand Americans across the country were asked to rank the honesty and ethical standards of various business and professional groups. Their response was that bankers, stock-brokers, and lawyers were lower in the minds of those participating in the poll than they had been in a similar survey in 1983 by the Gallup Poll.

The chairman of an organization which conducts seminars in business ethics says that violations of business ethics are resulting in a loss of public trust in business and the loss of billions of dollars. Examples cited are:

Officers' criminal acts caused the failure of more than one hundred banks and savings and loan institutions in 1985.

Large corporations have been caught making false charges that resulted in losses of millions of dollars to the U. S. government.

Employees stole more than $1 billion worth of merchandise from their employers in 1985.

The FBI estimates bank losses of nearly $300 million each year, much of it by computer.

His studies indicated that one of four employees is inclined to be dishonest. While persons in the companies he contacted said that "the ethical question is an extremely urgent matter, . . . where we probed their programs, we found next to nothing or very little by way of ensuring honesty or ethical behavior among their employees at all levels." (*Houston Chronicle*, May 19, 1986, Sec. 2, p. 1.)

It has been estimated that fraud costs the nation upwards of $200 billion each year and that "roughly one-half of the bank failures and one-quarter of the savings and loan collapses had as a major contributing factor criminal activities by insiders." (*New York Times*, Jan. 5, 1986, Sec. 3, p. 2.)

A recent survey of controllers on business ethics showed that businessmen are becoming more inclined to bend ethical rules to gain profits. Among the practices referred to in the survey were price fixing, insider trading, "greenmail," and recalculating or otherwise slanting the books to make the company or one of its units look better. One of the controllers responded, "We as business managers seem to be following the accepted norm today, which is, 'Let's see how much we can get away with.' The entire country seems to be preoccupied with finding ways around the ethical standards rather than abiding by them." (*Houston Post*, Nov. 24, 1985, p. E6.)

Federal Reserve Chairman Paul Volcker recently said that he worries about current business ethics. He stated, "I do have a feeling that people are willing to sail out closer to the wind these days than they were 20-30 years ago when I was in business." (*Houston Post*, Sept. 21, 1986, p. E9.) Increasing evidence of this is the fact that in almost every newspaper or magazine you read today there is an account or allegation of some questionable action in which companies and executives, frequently prominent ones, have been involved.

The following is a very perceptive analysis of "an era in which the cutting of ethical corners is deemed sound policy" and "where one ends up matters far more than how one got there":

> "[T]hose torn between the need to be true to themselves and the need for external validation have never had quite so much reason to feel isolated as they do today, for we are in a time of strikingly jarring contradictions. Never, on the one hand, has there been more ardent talk about resurgent morality, about a return to the tradi-

tional values of hearth and home and the merit of honest industry; and yet on the other hand, at once obscured and augmented by the ceaseless celebration of the success ethic, never has the definition of legitimate behavior in the world at large been so broad, or individual rigor been in such short supply. More and more unabashedly self-concerned, less and less committed to the ideals of altruism or the communal good, or even to intellectual honesty, we Americans lately find ourselves, by the millions, able to get around the best within ourselves as neatly as we get around the tax laws.

"This is not to suggest that we are innately short on character. To the contrary, we have shown ourselves in other, more morally demanding times to be as generous-spirited a people as any. But we live in quite the opposite of a vacuum—and so very much of what is coming our way just now drives home the same enticing message: if the bottom line is right, go for it." (Stein, "The Struggle Not to Sell Out," *Esquire*, Sept. 1985, p. 35.)

A comparable analysis was made over fifty years ago by Ivan Lee Holt, a Methodist bishop, who wrote:

"More people than ever before believe that compromise is essential to success and happiness; one must not be too good or too honest; principles must be surrendered if they seem to hinder advance. Those who reason thus can point to many in high positions who have reached the eminence by shrewd and dishonest practices. . . .

"We take our ethics as we take our athletics— watching other men play the game, to commend now, and to criticise later when there is a failure to give the best or to abide by the rules. A sense of obligation has slipped from our shoulders. We have wanted the fruits of civilization, but none of its burdens. Downright honesty we must have, or else our whole social structure collapses." (*The Return of Spring to Man's Soul*, Harper & Bros., 1934, p. 2.)

Ethical dilemmas confronting various professions, such as law, medicine, and accountancy, have required the adoption of codes of professional ethics or responsibility. These codes have taken various forms, depending upon the profession involved and the ethical problems which existed or were perceived to exist therein. While these codes are considered essential to the preservation of the integrity of these professions, they are too frequently violated. Too frequently also, the disciplinary bodies set up to enforce these codes do not act with sufficient dispatch or severity in dealing with the violators. Those violations not only have resulted in discredit and public criticism but also have subjected other persons to risk and injury.

Members of the professions should and must do a better job of policing their members. They must overcome a reluctance to report, investigate, present testimony against, and discipline their colleagues who engage in wrongdoing or who do not have the required level of competency. If the professions do not put their respective houses in order, the resulting criticism and problems will require that this be done through the legislative or regulatory process.

Business executives and managers also have been urged to consider themselves as professionals and to adopt a code of professional conduct. While this is a more difficult thing to accomplish because of the variety of businesses, positions, and ethical problems involved, some fundamental rules have been proposed. It has been said, "Ethical thinking is ultimately no more than considering oneself and one's company as citizens of the business community and of the larger society, with some concern for the well-being of others and—the mirror image of this—respect for oneself and one's character." (R. Solomon and K. Hanson, *It's Good Business*, Atheneum, 1985, p. 49.)

The following have been described as "eight crucial rules for ethical thinking in business" (*Ibid.*, pp. 45-49):

Rule No. 1: Consider other people's well-being, including the well-being of nonparticipants.

Rule No. 2: Think as a member of the business community and not as an isolated individual.

Rule No. 3: Obey, but do not depend solely on, the law.

Rule No. 4: Think of yourself—and your company—as part of society.

Rule No. 5: Obey moral rules.

Rule No. 6: Think objectively.

Rule No. 7: Ask the question, "What sort of person would do such a thing?"

Rule No. 8: Respect the customs of others, but not at the expense of your own ethics.

These rules can and should be followed by persons in all businesses. Top executives should set the example by observing these rules and insist that those under them do so. It is not only right but also "it's good business."

Companies have adopted codes of conduct dealing with conflicts of interest of officers and employees, and have adopted standards of business conduct. Also, many companies have compliance programs designed to avoid acts or practices which may be in, or may lead to, violation of the law. Again, it is important that these codes and programs be enforced and that those who violate them be dealt with properly.

A Stanford University professor who has conducted ethics workshops for companies has said that companies must overcome an assumption by many employees that top management wants profits however they have to be achieved. (*Wall Street Journal*, Sept. 9, 1986, p. 1.) While this assumption may not be a valid one in many companies, the pressure in some companies to increase profits is so great that the top management, whether they intend to do so or not, cause those under them to engage in improper and sometimes unlawful conduct to achieve management profit goals in order to advance or to keep their jobs. It is essential that corporate management communicate to their employees by their own actions and in writing that unlawful, unethical, and immoral actions in the conduct or furtherance of the company's business are not expected or condoned, will not be tolerated, and may be grounds for dismissal or other disciplinary action.

Another matter in the corporate area viewed by some as an ethical or moral issue is that of improving corporate accountability by changing the traditional modes of corporate governance. Much of the emphasis has been directed to the manner in which boards of directors should be structured and to defining the responsibilities of those boards in relation to the management of the corporations, the shareholders, and compliance by corporations with the law. (Letts, *Corporate Governance: A Different Slant*, 35 Bus. Law., July 1980, 1505.)

Corporate accountability was the subject of a study by the Securities and Exchange Commission over several years. (Kripke, *The SEC, Corporate Governance, and the Real Issues*, 36 Bus. Law., Jan. 1981, 173.) The American Law Institute also considered this matter. However, as is the history of many issues, the clamor and furor once associated with it seems to have died down. Perhaps one of the reasons is that many corporations have made changes in their boards as a result of some of the reform measures proposed.

These changes generally have been for the purpose of making directors more independent of management and more responsive to the interests of shareholders. Also, the growing risk of personal liability of directors if they simply rubber-stamp everything that management wants to do has had some effect on the attitudes and actions of directors.

Directors and officers of corporations generally are protected from personal liability for their business judgments by the business-judgment rule. That rule presumes "that in making a business decision, the directors of a corporation act on an informed basis, in good faith, and in the honest belief that the action taken was in the best interests of the company." (*Aronson v. Lewis*, 473 A.2d 805, 812 [Del. 1984].) However, the increase in the number of cases charging fraud, bad faith, or self-dealing by the corporate decisionmakers suggests "that the courts' traditional hands-off posture when faced with reviewing business decisions is giving way to demands for more protection for workers and shareholders." (Spiegel, *The Liability of Corporate Officers*, 71. A.B.A.J. 48, Nov. 1985.)

In recent years, we have seen accelerated takeover activity which "has focused attention on the legal, moral and practical questions faced by the directors of a company that becomes the target of an unsolicited takeover bid." (Lipton, *Takeover Bids in the Target's Boardroom*, 35 Bus. Law., Nov. 1979, 101.) The ethical and moral issues involved in takeovers and related matters—insider trading, "greenmail," "golden parachutes," and management buyouts—have been the subject of extensive debate.

There is the view expressed by some that the "extraordinary wave of takeover attempts in this country and the ingenious defenses erected against them" have resulted in a "'brain drain' diverting us from the development of new products, technologies and jobs, and focusing instead on the quick and easy profits to be made by corporate takeovers. We are putting our energies not into the long-term health of the nation but into short-run gains. . . . While great wealth is exchanged in such transactions, none is produced. Debt formation is invoked; capital formation is out of style." (Minow and Sawyier, "The Free-Market Blather Behind Takeovers," *New York Times*, Dec. 10, 1985, p. A31.)

The policy issue with respect to takeovers has been posed as follows: "Whether the long-term interests of the nation's corporate system and economy should by jeopardized in order to benefit speculators interested not in the vitality and continued existence of the business enterprise in which they have bought shares, but only in a quick profit on the sale of those shares?" (Lipton, *supra*, 104.)

Many say the hostile takeover phenomenon

"creates a market atmosphere of short-term gambling rather than long-term investing. It breeds a climate of fear, with hardly any companies safe from attack. Looking over their shoulders, managers focus on short-term results and defense mechanisms that often place additional high burdens of long-term debt on their companies. High leveraging not only makes companies more vulnerable to recession and high interest rates, but also undercuts product development, innovation and international competitiveness." (*New York Times*, Dec. 22, 1985, Sec. 3, p. 2.)

The Business Roundtable supports federal legislation "to insure that only those bidders who truly believe they can run a target more efficiently, and are willing to pay a premium for control would make tender offers." (*Ibid.*)

The contrary view espoused by those who support takeovers is that "cash tender offers increase everyone's welfare, and that the prospect of tender offers leads to more efficient management whether or not a bid is made for any given company." They contend "that any attempt by the target's managers to frustrate a tender offer should be prohibited." (Easterbrook and Fischel, *Takeover Bids, Defensive Tactics, and Shareholders' Welfare*, 36 Bus. Law., July 1981, 1733.)

In further support of this position, it has been argued that "takeovers are beneficial to both shareholders and society," that "defensive strategies designed to prevent takeovers do not improve shareholders' welfare," and that "the costs of defensive tactics and the costs incurred by bidders to overcome defensive tactics are sheer waste." (Easterbrook and Fischel, *Is Takeover Defense in Shareholders' Best Interest?*, Legal Times, Aug. 10, 1981, 42; and *The Proper Role of a Target's Management in Responding to a Tender Offer*, 94 Harv. L. Rev. 1161, 1184 [1981].)

It is my opinion that the damages done by the takeover mania in this country far outweigh any benefits claimed to be derived from takeovers. The reasons for that opinion are as follows:

1. While the management of some companies undoubtedly has been inept, has perpetuated itself, and has rewarded itself with excessive compensation and benefits, the solution to this is not a hostile takeover. There is no concrete evidence that the successor management will correct these conditions. Mismanagement and abuses by management can and should be eliminated by independent outside directors who truly are representatives of the shareholders and not cronies and rubber stamps of the corporate management. They should and must be persons of integrity who have the knowledge and capability required to determine whether management is doing a good job, and who are prepared to take action to remove corporate officers who do not perform in an acceptable manner. Directors who do not carry out their responsibilities to the shareholders should not be reelected by the shareholders. If the shareholders are not satisfied with the management of the company, they should vote against directors who allow the mismanagement to continue. Effective action by the shareholders may require that more information be provided to shareholders and that voting by shareholders be by secret ballot, a topic of current debate.

2. While hostile takeovers have driven up the price of the stock of the companies involved and in many cases resulted in substantial profits to share-

holders, they also have left in their wake many casualties. There has been "a tremendous human cost in takeovers. Men and women who have worked diligently for thirty and forty years are summarily dismissed. The cost to the former officer's [and employee's] self-esteem and to his family is immense." (*New York Times*, Sept. 7, 1986, Sec. 11 NJ, p. 34.) There also has been a cost to the companies involved and to the nation in the loss of experienced, productive persons whose knowledge and skills constitute a valuable asset.

3. Takeovers and the threat of takeover have diverted management from their responsibilities in planning and operations and have required them instead to devote their time and efforts to determining position and strategy in response to the takeover attempt. This diversion of time, talent, and energy is having an adverse effect on the ability of companies to compete in the marketplace, and its adverse effects on this nation's industry and economy may be felt for years to come.

4. Takeovers and the threat of takeover have caused both the target and the acquirer to take actions they would not otherwise have taken. Many of these actions have been detrimental to the company and the majority of its shareholders. They include restructuring, selling off business, new and excessive debt, raiding pension funds, issuing new classes of stock, adopting "poison pills," granting "lockup options," and issuing "junk bonds."

Takeovers and defensive actions against them not only raise issues of national and economic policy but also spawn numerous related issues presenting ethical considerations. "Ethics has been a concern ever since takeovers became big business a decade ago. The tactics of the trade have gotten nastier as the size and frequency of the deals have increased—with both the hunters and the hunted resorting to hard-edged ploys, all legal but some questionable on economic and moral grounds." (Pauley, "Merger Ethics, Anyone?" *Newsweek*, Dec. 9, 1985, p. 46.)

Some of the actions spawned by takeovers which involve ethical considerations are "greenmail," "golden parachutes," insider trading, and management buyouts.

"Greenmail" is the purchase by a company of its stock from a person or entity threatening to take it over at a price in excess of the market price and in excess of the amount the company is willing to pay other shareholders for its stock. The purpose of greenmail is to remove the threat of a takeover by the minority shareholder, who agrees not to pursue or renew the threatened action for an agreed period. In my opinion, greenmail involves ethical considerations on the part of both the company and the minority shareholder and is a practice that should not be engaged in by either. The minority shareholder involved in greenmail usually is someone who has recently acquired a large block of the company's stock for the purpose either of launching an attempt to acquire control of the company or to threaten the company with a takeover if it does not buy his stock at a premium. The management of the company in responding to that action by the payment of greenmail is doing so at the expense of other shareholders and, at times, is doing so to remove a "potential threat to their jobs." (*Wall Street Journal*, Oct. 21, 1986, p. 4.)

"Golden parachutes" involve severance agreements between a company and its top executives in which the company agrees to pay additional compensation to those executives in the event of a merger resulting in the loss of their jobs. Their original intent was to "put management in a neutral mode when recommending bids for the company" (Pauley, "Merger Ethics," p. 26) to prevent top executives involved in takeover fights from putting themselves before their shareholders. (Prokesch, "Too Much Gold in the Parachutes?" *New York Times*, Jan. 26, 1986, Sec. 3, p. 1.) However, golden parachutes have at times involved amounts so excessive as to defeat their purpose by causing top executives to support the merger to reap the benefits of their golden parachutes. They have been said to represent a breach of fiduciary responsibility and a waste of company assets. (*New York Times*, April 1, 1986, p. D31.)

The ethical questions and abuses involved in golden parachutes caused Congress to include in the Deficit Reduction Act of 1984 a provision to curb excessive golden parachutes. The act applies to golden parachutes entered into or amended after June 14, 1984. It defines as excessive those agreements providing for payments which equal or exceed three times what an executive's pay had averaged in the previous five years. It imposes substantial tax penalties on executives receiving more than allowed and on compa-

nies granting excessive payments. (Deficit Reduction
Act of 1984, Publ. L. No. 98-369, Sec. 67, 98 Stat.
494, 585-587 [1984]; Prokesch, "Too Much Gold.")
However, some companies recently have included in
the agreements provisions for compensating the executives for any tax penalties they may have to pay as
a result of receiving excessive severance payments.

Insider trading involves the use by corporate insiders or others of corporate information which is un-
available to the investing public in the purchase of
stock or options. In the "Doonesbury" comic strip of
August 10, 1986, a "leading investment banker" is
shown on television discussing the question, "Is there
a place for ethics in today's business climate?" In
answering that question "Yes," he shows a "typical
Wall Street meeting, filmed on a hidden camera," involving insider trading, in which the following con-
versation takes place:

> "Jim, I need some inside info on the Reamco
> takeover. Could be $250,000 in it for you, guy."
> "No can do, Stan. That information is confiden-
> tial."
> "How about $400,000?"
> "You don't understand, Stanley, it's illegal. More
> importantly, it's wrong. I just couldn't live with it."
> "Okay, $500,000."
> "Deal."

The commentator then states: "Ethics: a powerful
negotiating tool!"

The concept of insider trading being wrong origi-
nated on the basis of fiduciary duties of insiders, but it

has been broadened to include others obtaining possession of that information "premised on the notion that all investors trade equally and none should have an informational advantage over the other." (Farley, *A Current Look at the Law of Insider Trading*, 39 Bus. Law., Aug. 1984, 1771, 1773.) "Thus, anyone in possession of material inside information must either disclose it to the investing public, or, if he is disabled from disclosing it in order to protect a corporate confidence, or if he chooses not to do so, must abstain from trading in or recommending the securities concerned while such inside information remains undisclosed." (*SEC v. Texas Gulf Sulphur Co.*, 401 F.2d 833, 848 [2d Cir. 1968], cert. denied, 394 U.S. 976 [1969].)

While insider trading is not limited to takeover situations, "signs are that insider trading has become epidemic. Almost invariably the stock price of a company rises dramatically—and trading intensifies—in the days and weeks before a takeover is announced." (Lewin, "The Dilemma of Insider Trading," *New York Times*, July 21, 1986, p. D1.) Within the past five years, more than fifty people have been charged by federal prosecutors in New York with insider trading. These have included brokers, investment bankers, securities lawyers, and Wall Street market professionals, among others. (*Ibid.*)

The Insider Trading Sanctions Act of 1984 allows the federal government to collect a penalty of three times the amount gained or the loss avoided by insider trading.

Management buyouts also involve insiders and are situations where the management buys back the

company from the stockholders and "takes it private."
While the offer is more than the stock-market price, a
buyout presents an ethical question because of the
fiduciary relationship between the management and
the stockholders. The management has access to un-
limited inside information concerning the company
and the value of its assets. As a result, it is in the best
position to know what the price of the stock should
be, based upon the value of the corporation's assets
and its potential profits. The management obviously
must plan to make much more from the corporate
assets than they pay the stockholders in a buyout.
Otherwise, they would not incur the indebtedness
and risks involved. Therefore, there are sound
reasons for the position that a buyout is "unethical"
because the managers are violating their fiduciary
duties to the stockholders, are trading on inside information, and are not making the requisite full disclosure of every material fact in the solicitation.
(Stein, "Going Private Is Unethical," *Fortune*, Nov.
11, 1985, p. 169.)

Management at many companies in going private
has "abandoned traditional management functions in
favor of becoming asset price/stock price arbitragers"
and "corporate raiders." (Stein, "Buyout or Sellout,"
Barron's, October 6, 1986, p. 6.) As stated in a recent
analysis of the problem of management buyouts:

> "The basic problem is that management of
> publicly held companies are not supposed to be
> corporate raiders against their own stockholders.
> Officers and directors of publicly traded companies
> are not supposed to be arbitragers adopting an

adversarial stance against their own stockholders. Managers are not supposed to muscle stockholders out of the way when they see an opportunity to make money for themselves from the sale or development of corporate assets that, in the final analysis, belong to the stockholders—at least in legal contemplation.

"Officers and managers of publicly traded companies are supposed to manage the assets for the benefit of their employers, their wards, their stockholders. That is what being a fiduciary means and also, at least until fairly recently, what being a corporate officer meant." (Ibid., p. 7.)

While there may be valid reasons in some cases to take a company private, when this is done the management has a fiduciary responsibility to the stockholders to have a completely independent appraisal made of the assets and the potential profits of the corporation, to offer a price that represents full value for the stock, and to make the fullest possible disclosure to the board of directors and the stockholders.

"Ethical investing" is a growing practice in which American investors support their causes or social goals by their investments. According to the Social Investment Forum, which tracks this trend, "[m]ore than $100 billion in securities is now being 'socially invested', up from $40 billion in April 1985. . . . Seventeen states and sixty cities and nine counties have enacted legislation to prohibit pension funds, for example, investing in companies that do business in South Africa." (Myler, "Ethics Can Turn Out to Be a

Sound Investment," *Houston Chronicle*, October 7, 1986, Sec. 7, p. 1.)

The methods used by "ethical investors" involve: (1) not buying stocks in companies whose products they consider to be unnecessary or harmful or whose operations they find distasteful; (2) buying stock of those companies that have operations or objectives they consider to be good; and (3) launching a shareholder fight or proxy battle for a change in corporate policy. (*Ibid.*)

Understandably, there are differences of opinion as to whether the views and objectives of particular "socially conscious" investors are correct. However, there can be no question as to their right to press those views and seek to further those objectives through investments. Also, there can be no question about the fact that ethical investing is beginning to have some impact on the management of the companies whose products, operations, or objectives are involved. Ethical investing in some cases has not only been effective but also has been profitable. (*Time*, Oct. 27, 1978, p. 74.)

From the foregoing we see that there are many ethical or moral issues with which those in the professions and in business must deal in today's world. Many of these issues are intertwined with basic principles of morality and standards of conduct involved in everyday living.

A biblical saying familiar to all of us is "Do unto others as you would have them do unto you." A similar precept is also found in Confucian literature, where in response to a request from a disciple for a single word that could serve as a guide to conduct for one's life, Confucius is recorded as replying: "Is not reciprocity such a word? What you do not want done to yourself, do not do to others."

In our professional, business, and personal lives, each of us should conduct ourselves in accordance with the highest standards of ethics and morality and treat others as we would wish to be treated. By so doing, we can help make the Age of Technology one in which there is a strengthening of the character of our people, which has been and will continue to be a vital part of our nation's greatness.

Thomas Huxley, in his essay "Evolution and Ethics," wrote:

> "[T]he practice of that which is ethically best—what we call goodness or virtue—involves a course of conduct which, in all respects, is opposed to that which leads to success in the cosmic struggle for existence. In place of ruthless self-assertion it demands self-restraint; in place of thrusting aside, or treading down, all competitors, it requires that the individual shall not merely respect but shall help his fellows; its influence is directed, not so much to the survival of the fittest, as to the fitting of as many as possible to survive. It repudiates the gladiatorial theory of existence. It demands that each man who enters into the enjoyment of the advantages of the polity shall be mindful of his debt to those who have laboriously constructed it; and shall take heed that no act of his weakens the fabric in which he is permitted to live. Laws and moral precepts are directed to the end of curbing the cosmic process and reminding

the individual of his duty to the community, to the protection and influence of which he owes, if not existence itself, at least the life of something better than a brutal savage." (*Evolution and Ethics and Other Essays*, AMS Press, 1896, pp. 81-82.)

This concept is timeless and is as valid in this Age of Technology in which we live today as it was when these words were written nearly one hundred years ago.

THE MANY FACES OF TECHNOLOGY, THE MANY VOICES OF TRADITION

by

Martin E. Marty

Martin E. Marty

Martin E. Marty is the Fairfax M. Cone Distinguished Service Professor of the History of Modern Christianity at The University of Chicago, where he earned his Ph.D. in 1956. Since 1963 he has taught in the Divinity School, the Committee on the History of Culture, and the History Department.

Dr. Marty is Senior Editor of the weekly The Christian Century, *editor of the fortnightly newsletter* Context, *and co-editor of the quarterly* Church History. *He is President of the new Park Ridge Center, an institute for the study of health, faith, and ethics.*

Marty served ten years in the pastorate after having been ordained as a Lutheran to the Christian ministry in 1952. He is the Vice-President of the American Academy of Religion (to be President through 1988) and is Past President of both the American Society of Church History and the American Catholic Historical Association. The holder of 24 honorary doctorates, he is an elected Fellow of the American Academy of Arts and Sciences and an elected member of the Society of American Historians and the American Antiquarian Society.

The author of over thirty books, Dr. Marty won the National Book Award in 1972 for Righteous Empire. *His most recent work is* Pilgrims in Their Own Land: 500 Years of Religion in America *and this year the University of Chicago Press will publish the first of a four volume work,* Modern American Religion, *titled* The Irony of It All.

THE MANY FACES OF TECHNOLOGY, THE MANY VOICES OF TRADITION

by

Martin E. Marty

Technology wears many faces, one of them being dehumanizing and another being humanizing.

Some years ago the donors of a new center of communications assembled a cast of speakers to dedicate it. They, we, were all asked to comment on the role of technology in "the decline of the West." The choice of conference theme represented either an act of courage or of masochism, for the donors made their wealth from communications technology, which critics often score for contributing to the decline of the West.

Speaker after speaker failed to be unambiguously gloomy. They tended to be rather cheerful types, one of whom brightened the conference after he heard some doom-filled prophecies with the comment: "We do not know enough about the future to be absolutely pessimistic."

When some speakers pronounced forthcoming doom because of the burgeoning computer movement, atop the ever expanding and encroaching system of electronic communications, a sage among us rose to the occasion. A historian, he knew of a similar speech about one hundred years ago. It was given at the Philadelphia Exposition celebrating the United

131

States centennial in 1876. That speech, uttering similar words of doom about dehumanization, was directed against—pause—the telephone! Yet we survived. Of course, many of us hate the technology of the telephone that can interrupt our peace and solitude. But, said the intervener about the other, humanizing, face of technology: think how it helps people make contact if they are in senior citizens' homes, cancer wards, far from their own homes.

Let that incident be a parable for my topic and thesis. While some prophets of hope are Pollyannas who see nothing but promise in electronic technology and others are doom sayers who act as if we could or should get rid of technology if we would, I would pose the issue thus: in the search for the human, for values, for moral values, technology—to use a gambler's term—"raises the stakes on both sides." The question becomes now, "Is a humane technological order possible?" If so, what kind of critique must we make of the present? What might citizens do about various future prospects?

Technology Disrupts Tradition

Let us assume that human nature, whatever that is, does not change; what matters is the human in culture. I know of no evidence to suggest that a first-century person plopped into twentieth-century life would be a better or worse bundle of possibilities than we twentieth-century people. Yet we know that it makes a good deal of difference whether a person gets to live out her possibilities in, say, Albania, Inuit country of Canada, St. Tropez, or Dallas. Culture matters, and technology is a major instrument of culture.

Must we define technology? We cannot, of course, but we know it when we see it. We associate it with practical applications of science in the human world, with "technique" and what accompanies it. We see machines making machines, devices that help transform the natural world. To some it comes with ideology that assaults other ways of life; in the eyes of others, it is "neutral," and serves merely to render more efficient (or less efficient, say some victims of bad technology) processes that would exist in any case. In other words, for them television signals simply do better what smoke signals or messages carried by runners used to do.

For the moment we need only say that whatever technology is and however it does it, when this comes to communications and its corollaries, technology disrupts tradition. One historian speaks of "the world we have lost." The pre- or less-technological world made it possible for people to live in insulated and isolated cultures. Then came the wheel—and they could leave home. Then came the pump—and women no longer gathered at the well, there to gossip in support of traditional ways and values. Then came the whole industrial revolution—and people moved from town and country to city, where they met strangers and became victims of pluralism and, eventually, of relativizing pluralism.

The process goes on. Technological medicine brought by missionaries disrupts traditional medicine and, while it heals, can also cause people to forget some beneficial agencies and methods or understandings of suffering and well-being. When television came along it was clear to all that it might become pervasive and would disrupt insulated traditions; the television set in the monastery was an example. Fundamentalist colleges whose administrations forbade students from attending good new movies could not prevent them from seeing bad old ones in dormitories, thanks to television. Illustrations come easily to mind: Technology disrupts tradition.

Moral Values Inevitably Connect with Tradition

We cannot speak about moral values without speaking of tradition. Values refer to preferences based on a combination of philosophy or world view and habitual practice. We do not invent the network of values wholesale and out of thin air. We keep adapting what is already present, what is already the product of humans in culture. Values mean that we "prefer" this or that end-state to another, this or that means of approaching it to others. These acts of preferring are grounded in our deepest commitments. These commitments come from peers, from parents, even from remote and not named ancestors.

Eugen Rosenstock-Huessy has said, for instance, that the "tribe" and not the family is the *couche*, the matrix or womb or source of values. The family transmits and adapts them and projects them to a new generation. But members of a family do not have time enough to invent them all. Tradition, which comes from *traditio* and means "handing down," hands down a repertory or repository of values which include options for moral choice. We act because there were a Plato and an Aristotle, a Moses and a Maimonides, a set of people who have taken care to preserve and publish codes of laws or the Bible, because grandparents and parents have handed down some notions about how to act. Technology disrupts the acts of handing down by intruding on the connecting systems, by bringing in alien signals.

Technology Wears Many Faces, Produces Many Effects

Because of its disruption of tradition, technology appears to many people to be simply destructive. Ivan Illich is well known for showing how it can have negative effects on schooling or medicine or values in the Two-Thirds World. Peter Berger and colleagues write of *The Homeless Mind* dislodged by technology. Jacques Ellul in *The Technological Society* sees only one face to technology, an almost demonically efficient one. He writes that "the power of technique, mysterious though scientific, which covers the whole earth with its methods of waves, wires, and paper, is to the technician the abstract idol which gives him a reason for living and even for joy." How better speak of the role of technology in the field of values than to speak of idolatry, of deepest meanings? (*The Technological Society*, Vintage, 1964, p. 144.)

First, in respect to our approach to the topic, technology and especially communications media are destructive because they bring to us a pluralist world. When American Christians sent missionaries to bad pagan Buddhists and Hindus, all seemed well. When the senders met and came to honor the values of good

and spiritually or morally profound Buddhists and
Hindus, they created a new problem. How seriously
should we take our source of values if others can do as
well or better out of other sources? People turn, or
may turn, relativist when radio, cinema, television,
cassette, and computer come to their "tribe."

We see illustrations of this, for instance, in
portrayals of Amish life, long ago in *Plain and Fancy*
or more recently in *Witness* and most often in court
cases. The Amish can isolate their children from alien
traditions in their elementary schools in rural
America. They cannot do so when these children
must go to more urban and pluralist consolidated high
schools. The pressure of peers shaped by mass media
will lure young Amish children to the sinful world of
cheerleaders in short skirts, young people who listen
to rock music, and the like. We onlookers see something admirable in what is disrupted and inevitably in
the disrupting (and possibly liberating) agent. Technology wears more faces than one.

So it is that today we have "textbook" battles over
signals that come from a pluralistic culture and violate
the tribe of fundamentalism. So we all complain that
"our" kind of family is never shown or honored on
prime television. The values we nurture are undercut
or jostled by others, and that is bad.

Alongside pluralism comes the observation that
modern technology of communication promotes the
secular. This disturbs because moral values for the
vast majority of citizens connect with explicitly
"sacred" signals, with traditions that connect us with
the transcendent, the divine, probably the divine as a

particular set of canons and codes and communities describe this. Television does not seem neutral; it comes as a destructive force. This is so—*pace* some critics—not because of a "secular humanist conspiracy" but because of a lack of memory and imagination among communicators and, even more, because of pluralism. What part of what tradition should the media regard as representative and normative? Will non-Mormons permit Mormon values to come into their homes? Will five million Mormons buy deodorant from communicators who provide non- or anti-Mormon spiritual signals? So communicators try to be silent. Yet there is no "ethical vacuum," as philosopher Hans Jonas and many others remind us. Technology must bring some signals from some system and, by default, this tends to be secular and thus competitive to most peoples' personal and tribal choices.

Thirdly, this means that technology seems or is subversive. This is why so many thoughtful philosophers and counselors in the liberal Western tradition, where the great "advances" are made, have so regularly dealt with one face of technology. It dehumanizes, depersonalizes, creates artificial and illusory cultures and values. Yet we can also hear that the subversion may have positive effects. I choose but one witness, European politician-engineer Egbert Schuurman. For him the spiritual and, indeed, Christian meaning of technology implies "emancipating the body and mind from toil and drudgery, repelling the onslaughts of nature, providing for man's material needs, and conquering diseases . . . eliminating unnecessary burdens, freeing time, promoting rest and peace. . . ." (*Reflections on the Technological Society*, Wedge, 1977, p. 21.)

Every reader can take a moment to imagine how Schuurman's positive scenario applies in the most threatening/promising sphere, television. Can we not think of moments of profound and delightful entertainment? Has not the picture of ethnic and race relations changed, often positively because the white majority sees affirming images of Asian and African people in serials, broadcasting the news, being in love, caring about human good? Don't we meet saints and heroes on television and don't we confront moral crises there? Think how close South Africa and Nobel prize winners become when the instrument of television invades our living room. We become informed and can make new choices. Indeed, "the stakes are raised on both sides," but that affirmation does include the underused and underviewed possibility that the raising is also on the side of promoting moral values.

Technology Brings Ambiguous-toward-Positive Effects

A survey of the impact of technology shows much in support of Schuurman's view, but never unambiguously. Ernest Gellner has reminded us that, over the long view, humans given the choice will choose longer life over shorter, more comfort over more pain, more meaningful work over drudgery, some social security over none. Hence they will take the mixed blessings of technology.

Given the choice, in medicine, most will accept the gifts of technological medicine even as they rue the loss of the personal touch and must recover the wholism that technology can disrupt. Most faith-healing religions are glad to turn to surgery or novocaine for support. We must worry about distribution of health care because of the costs of technological medicine, but most would rather worry about justice and fairness than have nothing to fight over in the first place.

The computer can inspire agony; ask anyone who has had a misunderstanding with a computer at a credit-card company. Yet the computer can also help people make connections, can help them plan strategically—often to good moral effect. It would be phony of me to write in praise of the yellow pencil, using my word-processor. Or to criticize the technology of printing while hoping that this book and others can be produced and reach publics.

So with the media. I recall once riding with the mayor of Beer Sheba in southern Israel. Outside the window there was a scene of Bedouin and thus Arab people tending sheep. It looked, I said, just like the pictures in my childhood "Bible history" books. Nothing had changed. I all but expected Isaac or Ishmael or Rebecca to step forward. "Yes, nothing looks different: the tents, the crook, the sheep, the garb are the same. Notice that young sheepherder, though. He is holding a transistor radio, and it is tuned to Cairo." We all know that the Ayatollah Khomeini effected a revolution against the West and technology from his "electronic fortress" in Paris. Thence he shipped cassettes to villagers who got their signals from his voice, on them. The consequence was inconveniencing to us in the West, but seemed liberating to those who participated in the revolution. Here, as often, technology had ambiguous kinds of effects, depending upon who was "liberated" and who set back by it.

The Situation Today

Today Americans speak of an intensified "values crisis." Under other terms, people have spoken thus since Adam and Eve were fifty years old. Cultural change brings a threat to traditionalists or custodians of tradition and hence of moral values. They do not often see that change can be for the better as well as for the worse, or do not remember that change for the better can be as inconveniencing as change for the worse. Yet there are good reasons for citizens today to talk about sudden paces of change and a crisis of values. We can survey some of the main areas of concern and trouble.

The family. Technology disrupts it. Some disturbance can come because marriage and sex counselors write on "open marriage." But more comes on more neutral grounds. We might as well blame Big Mac as Dr. Ruth for interruption of tradition. As both parents work outside the home, as children's school schedules conflict with meal time, as we choose a life with many extra-home commitments, it is easy to "graze" at fast-food outlets which exist because of technology. Without at least one common meal a day on most days—and most families do not have that now—there is little chance for appraising, learning, and transmitting traditions of moral values. So we strike at "secular humanists" who invade our home with televised undercutting signals.

Sexuality. The invention of the condom and then the pill, the devising of *in vitro* fertilization and the increase of possibilities for convenient abortion, all connect with technology. They bring changes that may be liberating to some, but strike terror in others. The media, for example cinema and the prime-time serial, constantly glamorize sexual values in conflict with those of traditional culture. So does the world of advertising.

Defense. Technology produces ever greater weapons of destruction and the threat to use them. Some "traditionalists" believe that all such weapon systems imaginable should be developed and poised. Others, such as the Roman Catholic bishops, deal with traditions of pacifism and "just war" and find that technological warfare transcends the bounds. What they discuss is brought into the homes of citizens via technologically produced newspapers and television. A values crisis results.

The classroom. The modern textbook reflects and is produced by technology and brings into the second- or third-most intimate zone of community life (at least the family used to be first) a set of signals that jostles those some parents would nurture. In the act of bringing a larger world the text and classroom compete with the chosen and cherished smaller world of fewer signals. Some parents react against "secularism" in the textbooks. We have a values crisis.

The media, however, are most crucially poised, and that is why I have been asked to address their sphere. I have to point to ambiguity here, by referring to a paradoxical situation. Even those who would resist technology of modern life may use it, may use it most efficiently. Thus in religion, while we "liberals"—I can think of no profound sage or counselor who has not done so, be he named Buber or Berdyaev or Tillich or whoever—worried about dehumanizing and depersonalizing technology, fundamentalists used radio and now television and mail communications to promote antimodernity.

One can confront pulpits with instruments for changing lighting, heat, and sound; they communicate whether a child cries in a nursery or a car has lights on in a parking lot. And then, from that pulpit, will come a sermon against technology and modernity. What goes on? Theorists of culture have often observed that late-comers—the fundamentalists are thus regarded—know "the privilege of historical backwardness" and "the merits of borrowing," to cite Thorstein Veblen and Leon Trotsky. That is, they are more free than the inventors to take selectively some elements of the invented world and freely use them to their purposes. This further illustrates that there can be no escape from technology, only a variety of responses and uses.

However, technology here does not come neutrally. The transformer is transformed, the broadcaster is changed, the receiver of communications walks into an altered world. To illustrate: the modern television evangelist/entertainer must attract viewers and financial supporters to stay in business. There are always better cameras and studios to buy and build in the name of the very technology that may be theoretically scorned in the "antimodernist" world of the evangelist. How does one hold an audience? "You're only as good as your last act." You have to be sensational, in order to outdo the competition and hold your audience from last week.

How attract? "Take up your cross and follow Jesus" may be heard, but it will not hold an audience. Once upon a time traditional evangelical cultures belonged to many of "the disinherited" or "disaffected." Religion offered hope in the life to come. Now "Dives" of Jesus' parable might be enjoying things while I, Lazarus, sit hungry at the gate. You will not often hear or see that story on competitive religious television. The most popular passage has become the word of Jesus in the Fourth Gospel: "I am come that you may have life and may have it more abundantly." Whatever he meant, he did not mean the material abundance dangled or promised those who are converted by, attracted to, and supportive of the electronic evangelizer. A values shift has gone on, one that brings people into a consumerist culture that denies many values of the Gospel tradition as the grandparents cherished it.

In our world of technological ambiguity, then, we see here, as so often, a double effect or possibility. The stakes are raised on both sides. What comes of "traditional" values? One of my students is writing on the traditional values of evangelical subcultures "then" and "now," with many changes resulting from technology. Then, fifty years ago, one was taught not to love the world, or at least to be suspicious of the sphere where moth eats and rust corrupts; one should lay up treasures in heaven. Women should not bob their hair or use cosmetics. They should not covet or be attracted to or possess expensive apparel. People should not seek worldly fame. The worlds of athletics and entertainment were competitors to the church.

Today it is precisely the world of electronic evangelism/entertainment that has changed most, has been changed more than it has changed the world. The athlete and beauty queen are honored as Born Again witnesses. Those who tithe to the program will receive double tithe benefits back from the Lord. Those who pray with the television healer will gain healing; one does not see the unhealed, the disabled, the ugly, the addicted. What happens to the cross in such a transit of values? As a fellow Christian at the borders of the evangelical camp—an "evangelical" but not in the clubby sense—I may applaud the fact that this cohort has come to honor the good in God's creation. I might admire the muscle of the athlete, the beauty of the woman, the vitality of the culture. But I cannot be convinced that a transvaluation of values has not gone on here, one that is part of the "values crisis" and induces ambiguous moral change. The stakes, again, have been raised on both sides.

I have been asked to dwell on this case study for a while, since so much of the "moral values" crisis talk comes from one multimillion member cohort of our society, the producers and receivers of electronic

Gospels. We have to notice, close up, some of the changes. This "electronic church" must live off of its constituency and clientele. A pastor must also live in relation to people in the pew, and may not always be free to be thoroughly prophetic, but the stakes have been raised here. There is distortion of the product, and there is illusion. The communicator suggests that television is a good instrument for conversion. Yet only one-tenth of 1 percent of adults who came to church life as adults in America credit radio-television-newspaper as the agent. Over 80 percent credit the witness and example of invitation of someone well known to the convert, be he spouse or she parent or they friends.

Such electronic entertainment has found a niche in a subculture. On the positive side, it ministers to shut-ins and the neglected. On the negative side, it accepts the world of celebrity with its trappings and tendencies toward idolatry or at least false attraction. Such media have a narcotic effect. Surveys find that the ten to twenty million people who are part of the electronic church watch television 5.5 hours per day—most of it, obviously, of the secular-pluralist "prime time" sort. The addiction may undercut other values. A recent responsible survey found that Christian Broadcasting Network was not the sexiest but was the most violent, insofar as its images projected were concerned. (It buys up old serials from television, and most are of the violent sort.) Yet violence here comes with Christian sanction and endorsement, making it "more O.K." Such communication creates an electronic envelope of values and expectations that jostle traditional signalers, as in the local congregation. "Why can't we have Fran Tarkenton or Loni Anderson witness as *our* church, the way they do on the electronic church or the congregations that live off of it?"

Most problematic from the "moral values" angle is the way it creates a subculture that gives only negative views of the larger culture. By now the reader knows me well enough to know that there can be a positive effect, too: This subculture has demonstrated proper dissatisfaction with The Culture and its often banal and evil signals. Yet there is here little opportunity to show God working in the society where people do not know God in one way. Jimmy Swaggart in 1986 endorsed the possible presidential candidacy of competitor-colleague Pat Robertson. He said that were Robertson elected, the hand that was on the Bible for oath of office would belong to someone whose head and heart and shoulder were right with God in Jesus Christ—and this would be the first time this would happen "in human history"! That revealing bit of hyperbole showed how even the "other" Christians' world, most "even" that of Born Again Jimmy Carter, did not count before God because it did not come under the right auspices.

One can concentrate too long on such a case study. Most Americans do not participate in the electronic church, and millions may get good effect from some of it. The reformer would find better places to start a crusade in our society. So some waste money through its false lures? So do more on the lottery, the horses, the football players who pay off less and less frequently. So some watch too many hours? At least it keeps the worst of them off the streets. And some of them no doubt have moral values reinforced and carry them out to a world that needs at least some of them.

Is a Humane Technological Order Possible?

After all the surveys and case studies we are left with our question. I recall one evening when two Chilean Jesuits were meeting with some of us North American theologians. Two of the theologians were "liberationists" who decried imperial technology and its disruption of Third World cultures. They romanticized the world of the Chilean peasant or slum dweller who was not jarred by television or corrupted by technological healing or advertising. Finally one of the Jesuits said, "No, no, dear colleague. Do not speak for us. We would like our people first to have your problems before they seek your solutions. First give them means and comfort and enjoyment before you see how it might bring bad changes—and *then* work for improvement." Another parable. Technology will not go away; it will advance. How humanize it?

It is possible to sketch some outlines. First, technology needs compensatory complementation and supplementation. This means, for instance, that alongside technological medicine there must be a conscious development of "care structures" or, better, the human touch. Alongside electronically propagated Gospels there must be ever richer community life in congregation. As technology comes, there must come buffers, partial insulators, interpreters.

Secondly, there have to be ever more strenuous

efforts to use persuasion, not coercion, to promote moral values. It is tempting to seek to be Big Brother, to envy those with mass communicative power, to link one's own system with the ballot box and impose values on people where there is no moral consensus. Such efforts overlook the positive values in subcultures and traditions. Thus the notion that all would be well if we would pass laws privileging and favoring "the Judaeo-Christian" tradition would solve little. Whose Judaeo-Christian tradition? Jews'? Christians'? Protestants'? Catholics'? Mormons'? Gentiles'? If Baptists', which Baptists? Such questions point to the fact that we have problems within pluralism apart from technology, and technology of communications brings them close to home but is not the sole agent of confusion.

What do we do?

1. We seek to interpret technology. To have a handle on what goes on is itself part of counteraction and therapy.

2. Citizens can work to understand the traditions of moral values. Eugene Goodheart says of those who reject what is handed down from, say, Greece and Rome or Jews and Christians: you may not possess the traditions, but the traditions possess you. Our actions are habitual, gestured, reflexive. What are their sources? What in those sources is of value?

3. People can engage in selective retrieval from the many voices of tradition. Much tradition, despite idolizations of it in our nostalgic culture today, was burdening, stultifying, static, enslaving. One must

learn to select out of what we inherit what can be of critical value in the present generation—and then project it, perhaps electronically!

4. Concerned folk should rebuild subcommunities. Margaret Mead said twenty years ago that because parents so often had no sense of tradition and no courage to try theirs out, the children were forced to make a mishmash of all the religions there ever were—and it didn't work and couldn't last. The subcommunities are somehow accessible in the technological world, and they include humane and humanizing elements.

5. In the republic of technology there must be concern for the republic of values. The subcommunities must interact, overlap, support each other. You cannot have a republic of sealed-off moral-values-generating-systems. Mormons and Episcopalians, secularists and Baptists share and must share some values and hopes in the larger culture. Mass media can promote their interaction.

6. There can be stress, if citizens care, on values that come with the larger value systems of our republic. We have shared and do share some common memory, narrative, myth, story, history. The Pilgrims and Einstein and Martin Luther King and war heroes and saints on this soil belong to us all, not just to subcommunities. We share common propositions of moral value ("We hold these truths . . .") and a common legal-constitutional system that implies more. We have common projects, problems, intentions, partial resolutions, and they need communicating and reinforcing. We might even have what

Jefferson called "common affection" or a kindred spirit—and technology need not take that away. It may even enhance it, if we intervene and insist that it be supportive.

Tradition, in a republic, has many voices. Technology has many faces. Human agents, many of us would say "under God," are not deprived of all choices in the presence of both. And that is a fairly cheerful or energizing prospect in the face of the doom that would result if we get it, or keep getting it, all wrong.

THE UNCHANGING SPIRIT
OF FREEDOM

by

Andrew R. Cecil

Andrew R. Cecil

Andrew R. Cecil is Distinguished Scholar in Residence at The University of Texas at Dallas and Chancellor Emeritus and Honorary Trustee of The Southwestern Legal Foundation.

Associated with the Foundation since 1958, Dr. Cecil helped guide its development of five educational centers that offer nationally and internationally recognized programs in advanced continuing education.

In February 1979 the University established in his honor the Andrew R. Cecil Lectures on Moral Values in a Free Society, and invited Dr. Cecil to deliver the first series of lectures in November 1979. The first annual proceedings were published as Dr. Cecil's book The Third Way: Enlightened Capitalism and the Search for a New Social Order, *which received an enthusiastic response. He also lectured in each subsequent series. Another,* The Foundations of a Free Society, *was published in 1983. A new book,* Three Sources of National Strength, *appeared in 1986.*

Educated in Europe and well launched on a career as a professor and practitioner in the fields of law and economics, Dr. Cecil resumed his academic career after World War II in Lima, Peru, at the University of San Marcos. After 1949, he was associated with the Methodist church-affiliated colleges and universities in the United States until he joined the Foundation. He is author of twelve books on the subjects of law and economics and of more than seventy articles on these subjects and on the philosophy of religion published in periodicals and anthologies.

A member of the American Society of International Law, of the American Branch of the International Law Association, and the American Judicature Society, Dr. Cecil has served on numerous commissions for the Methodist Church, and is a member of the Board of Trustees of the National Methodist Foundation for Christian Higher Education. In 1981 he was named an Honorary Rotarian.

THE UNCHANGING SPIRIT OF FREEDOM

by

Andrew R. Cecil

"The God who gave us life, gave us liberty at the
same time: the hand of force may destroy, but cannot disjoin them."
> Thomas Jefferson, *A Summary View of the
> Rights of British America*, July, 1774.

Even before our nation was born, Edmund Burke
remarked on March 22, 1775, in his "Speech on Mov-
ing His Resolutions for Conciliation with the Colonies" that "In the character of the Americans a love of
freedom is the predominating feature which marks
and distinguishes the whole." That observation by
Burke has continued to be true throughout the history
of the United States. During the ensuing two centuries, the world has undergone dramatic changes, but
the spirit of freedom that Edmund Burke saw in our
ancestors lives on in their descendants. Among all the
social, political, and technological changes that our
nation has undergone, that spirit of freedom has
proved to be the unchanging force that underlies all
that we have undertaken and accomplished.

The essential exponents of such a spirit of freedom
are free people. The citizens of our nation enjoy as
their birthright an unparalleled network of civil, constitutional, and human rights. Neither their govern-

ment nor their fellow citizens can deny or abridge these rights, and they can be limited only under carefully defined conditions. The American colonies were founded upon the premise that nothing is more valuable than the worth and dignity of each individual—which are derived from man's unique relationship to the Creator of the Universe. This faith in the greatness of the individual was expressed beautifully by Walt Whitman:

> "I swear I begin to see the meaning of these
> things! . . . it is not America who is so great,
> It is not I who am great, or to be great—it is you
> up there, or anyone;
> It is to walk rapidly through civilizations, govern-
> ments, theories,
> Through poems, pageants, shows, to form great
> individuals.
> Underneath all, individuals! . . .
> The American compact is altogether with indi-
> viduals. . . ."

James Madison wrote, "We rest all our political experiments on mankind's capacity for self-government." The phrase "self-government" means not the government of each by himself but of each by those who have succeeded in making themselves accepted as the majority. Limitation of the power of government and vigilance against the abuse of such power, however, lose none of their importance when those who hold power represent a majority. John Stuart Mill, the most widely known nineteenth-century British political philosopher, warned against

the danger of tyranny by the majority and stated in
On Liberty that there is also a need for protection
"against tyranny of the prevailing opinion and feeling,
against the tendency of society to impose, by other
means than civil penalties, its own ideas and practices
as rules of conduct on those who dissent from them."

The faith of the founders of the United States that
the Creator has endowed all human beings with the
right to be free and to be protected against the
possible tyranny of their government endowed our
nation with a sense of mission which was expressed in
the opening paragraph of the Federalist papers. By
our "conduct and example," it was said, our people
hoped to demonstrate to all the world the advantages
of a free and self-governing society. This mission became known as the "Great American Experiment."
All the world looked on to see whether a society
founded on a faith in self-government and a belief in
human dignity could survive.

Throughout the years, the rest of the world has
seen the result of the "Great American Experiment."
Freedom proved to be not simply an ideal, but a force
with magnetic powers to attract—sometimes against
very great odds—those who lacked it or lost it but
never stopped desiring it. In recent decades it has
attracted more than three million Germans, who have
gone from East to West; more than two hundred
thousand Hungarians, who escaped to freedom during the rebellion in 1956; and hundreds of thousands
of Vietnamese, Cambodians, and Laotians, who left
their countries behind and took the risk of being
killed or raped by modern-day pirates and marauders.
Escaping from deception, coercion, terrorism, and as-

sassination, they left their homelands, friends, families, and possessions. The pain and desperation that forced their escape can only be compared to the suffering of an animal that, in order to escape from captivity, gnaws off its own limb when that limb is caught in a trap.

Why did so many seek to escape their previous lives at such cost in order to seek freedom? They did so because human nature looks for a government which will discharge its responsibilities for the welfare of the nation and the individual without depriving the people of the right to participate freely in the formulation of laws and without the danger of dictatorship and tyranny. To avoid such a danger and to preserve a democratic state, it is imperative for each individual to preserve his sense of responsibility for the integrity of the government. Thomas Paine wrote, "Those who expect to reap the blessings of freedom must, like men, undergo the fatigue of supporting it." The price of freedom is eternal vigilance on the part of the citizen, and his participation, whenever an adequate opportunity arises, in the functions of government.

Freedom, according to Mill, comprises: first, liberty of conscience in the most comprehensive sense, liberty of thought, from which it is impossible to separate the cognate liberties of speech and the press, and absolute freedom of opinion on "all subjects, practical or speculative, scientific, moral, or theological"; secondly, liberty to form our lives to suit our tastes and characters, so long as what we do does not harm our fellowmen; and thirdly, freedom to unite for any purpose not involving harm to others.

The first ten amendments were added to our Constitution to preserve the precarious balance between liberty as self-government and liberty as the freedom of the individual from government. Within that framework came freedom of religion; freedom of speech and of the press; freedom of peaceable assembly; the right to keep and bear arms; the prohibition of peacetime quartering of troops without consent; the prohibition of more than one punishment or one trial for the same offense (double jeopardy); protection against self-incrimination (no person can be compelled to witness against himself); protection against loss of life, liberty, or property without due process of law; freedom from excessive bail and from unreasonable searches and seizures; and the right to speedy and public trial.

The full scope of the liberty guaranteed by the Constitution cannot be found in or limited by the precise terms of the specific guarantees provided in the Constitution. This liberty, wrote Justice Harlan, "is not a series of isolated points pricked out in terms of the taking of property; the freedom of speech, press and religion; . . . and so on. It is a rational continuum which, broadly speaking, includes a freedom from all substantive arbitrary impositions and purposeless restraints. . . ." (*Poe v. Ullman*, 367 U.S. 497, 543, 81 S.Ct. 1752, 1776 [1961].) The Constitution and Bill of Rights assert the existence of certain liberties of the American citizen that are a part of our government. But are these great documents the source of our freedom?

Learned Hand, the illustrious judge, once wrote: "I often wonder whether we do not rest our hopes too much upon constitutions, upon laws and upon courts. These are false hopes; believe me, these are false hopes. Liberty lies in the hearts of men and women; when it dies there, no constitution, no law, no court can even do much to help it. While it is there, it needs no constitution, no law, no court to save it." (Learned Hand, "Speech to Newly Naturalized Citizens," May 21, 1944, *The Spirit of Liberty,* ed. Irving Dilliard, New York, Alfred A. Knopf, Inc., 1952, p. 189.)

In the same spirit, Martin Luther long before observed that "there can be no better instructions in . . . all transactions in temporal goods than that every man who is to deal with his neighbor present to himself these commandments: 'What ye would that others should do unto you, do ye also unto them,' and 'Love thy neighbor as thyself.' If these were followed out, then everything would instruct and arrange itself; then no law books nor courts nor judicial actions would be required; all things would quietly and simply be set to rights, for everyone's heart and conscience would guide him."

Sources of Freedom

The American concept of freedom rests not "upon constitutions, upon laws and upon courts" but derives its character from three main qualities: religion, with its emphasis on morality; knowledge, with its emphasis on education; and the rule of law, rooted in common religious and moral tradition.

A. *Religion*

Religion as a source of freedom should be distinguished from the right to religious freedom that gives every human being the right to honor God according to the dictates of his conscience, and the right to worship privately and publicly. Thomas Jefferson's famous Virginia Bill for Religious Liberty illustrates this distinction. The preamble to the bill states among other things that "Almighty God hath created the mind free." This faith in God's being "Lord both of body and mind" is the source of freedom. The statute itself, which proclaims "that no man shall be compelled to frequent or support any religious worship, place, or ministry whatsoever, nor shall be enforced, restrained, molested, or burthened in his body or good, nor shall otherwise suffer on account of his religious opinions or belief," stresses the right to religious freedom, which in the United States is only one of the freedoms enjoyed under our Constitution and Bill of Rights.

With James Madison, Jefferson led the fight against a tax levy for the support of the established church in Virginia. They aroused the sentiment that culminated in the adoption of the Bill of Rights. Jefferson, in bitterly opposing the government-established religions, stressed "the firm conviction" in the minds of the American people that their freedom derives from God. Because of this "firm conviction," Alexis de Tocqueville described religion as the "foremost political institution." "In France," he wrote in *Democracy in America* (1839), "I had almost always seen the spirit of religion and the spirit of freedom marching in opposite directions. But in America I found they were intimately united and they reigned over the same country." Religion was, according to the Founding Fathers, the major source of our liberties and of the protection of the individual against the suppression of these liberties by a dictatorial government.

President Washington proclaimed November 26, 1789, a day of thanksgiving to offer "our prayers and supplications to the Great Lord and Ruler of Nations, and beseech Him to pardon our national and other transgressions. . . ." Almost all of the presidents who succeeded him proclaimed Thanksgiving, with all its religious overtones, a day of national celebration. References to our religious heritage are found in the statutorily prescribed national motto "In God We Trust" and in the Pledge of Allegiance to the American Flag, where we are "One nation under God." Chaplains of the Senate and the House and the military services are paid from public revenues. There are countless other illustrations of the acknowledgement of the role of religion in American life that support the opinion expressed by Justice Douglas, "We are a religious people whose institutions presuppose a Supreme Being" (*Zorach v. Clauson*, 343 U.S. 306, 313, 72 S.Ct. 679, 684 [1952]).

By asserting freedom of conscience as an indefeasible right, religion is one of the most powerful elements which enters into the formation of the concept of individual liberty. The principle of absolute freedom of conscience denies that an individual is accountable to others for his religious beliefs. Man's belief in a superhuman power which directs and controls the course of nature and human life and the bond uniting him to God should not be confused with dictates of organized religion or with religious practices.

History shows that some activities of the organized church, claimed to be undertaken to preserve purity of faith and morals, brought disgrace and were viewed with terror. The Fourth Lateran Council of the Roman Catholic Church convened in 1215 by Pope Innocent III decreed that Jews should wear clothing to make them distinguishable from Christians, should not appear in public during Holy Week or Easter, and should not hold any public office where they might exercise authority over Christians.

The Inquisition, apparently instituted in the thirteenth century, and especially the dreaded Spanish Inquisition, interested chiefly in the faith of the converts from Islam and Judaism, permitted torture, including burning at the stake, in the examination of those accused of heresy as well as of the witnesses. There was widespread confiscation of the property of the condemned heretic by the heresy-hunters. The far-reaching abuses included the questioning of Saint Ignatius of Loyola, the founder of the Society of Jesus and a leader in the Catholic Reform, for suspected heresy.

Following the Reformation thousands of Christians were massacred in Europe only because their religious convictions differed from those that prevailed in the countries where they lived. The U. S. Supreme Court, reviewing the background and environment in which the First Amendment was adopted, pointed out that the centuries immediately before and contemporaneous with the colonization of America were filled with turmoil and civil strife generated in large part by the established sects determined to maintain their absolute political and religious supremacy through persecution and oppression.

With the power of government supporting them, stated the Court, at various places and times Catholics had persecuted Protestants, Protestants had persecuted Catholics, Protestant sects had persecuted other Protestant sects, Catholics of one shade of belief had persecuted Catholics of another shade of belief, and all of them had from time to time persecuted Jews. In order to force loyalty to the religious group dominant at a particular time and place, men had been cast in jail, tortured, and killed for expression of disbelief in church doctrines, for disrespect for government-established churches, for nonattendance at the churches, and for failure to pay taxes and tithes to support them. (*Arch R. Everson v. Board of Education of the Township of Ewing*, 330 U.S. 1, 67 S.Ct. 504 [1947].)

This religious intolerance and fanaticism prompted Christians in Europe whose faith deviated from the dominating "true faith" to flee to America. They came to this country to escape the bondage of laws which compelled them to support and attend government-favored churches. The freedom-loving settlers were therefore shocked when they faced the abhorrent practice of imposing taxes on believers and nonbelievers alike to build and maintain churches and church property and to pay ministers' salaries, transplanted to and thriving in the soil of the new America.

Cruel persecutions, the inevitable result of government-established religions, were a repetition of the Old World practices. Men and women of varied faiths who happened to be in a minority in a particular time or locality were persecuted because they persisted in worshipping God the way their own consciences dictated. Catholics found themselves hounded, Quakers were imprisoned, and because of this persecution in the Massachusetts Bay Colony Roger Williams was banished from the Colony in 1635. The founder of Rhode Island, accompanied by four others, established a settlement called Providence, which England granted absolute liberty of conscience in religion.

This spirit of liberty was not universal among the founders of our nation. Tolerance was not one of the virtues of the Puritans in England or in the United States. They considered, for instance, Christmas Day as "the old heathen's feasting day in honor of Saturn," as "a popish frivolity," and as the "dreadful work of Satan." Plymouth's Governor William Bradford (1590-1657), who adopted a more liberal attitude to those who belonged to other religious groups than did the Puritan leaders at Boston, reprimanded the colonist who took Christmas Day off "to pitch ye barr, and play at stoole ball and such like sport." Since the Catholic church celebrated Christmas Day, the purpose of the ban on frivolity on this day was "to prevent the observing of such festivals as were superstitiously kept in other countries to the great dishonor of God and the offense of others." Until 1856 the children in the area of New England had to attend schools on Christmas Day.

When charters granted by the English Crown authorized individuals and companies to erect

religious establishments, and when almost every
colony exacted some kind of tax for government-
sponsored churches whose ministers preached
sermons generating a burning hatred against dis-
senters, the climate was ready for a movement to strip
the government of all power to tax, support, or otherwise assist any or all religions, or to interfere with the
belief of any religious individual or group. This
movement explains the enactment of the First
Amendment.

Religion as a source of our freedoms should not be
confused with the attempts of organized religion to
give direction in politics from the pulpit, or with
claims by politicians that they have a special relationship with God. In the heat of political controversies,
those who challenge the views of a particular group
are looked upon as "subversive liberals" or "reac-
tionary conservatives." The fundamental principles of
the Church become obscured by the belief that some
economic or political point of view is ordained by
God, and, therefore, adherence to this point of view
is construed as a test of loyalty to the Church.

As the Supreme Court has stated, political debate
and decision, however vigorous or even partisan, are
a normal and healthy manifestation of our democratic
system of government, "but political decision along
religious lines was one of the *principal evils* against
which the First Amendment was intended to protect. . . . The history of many countries attests to the
hazards of religion's intruding into the political arena
or of political power intruding into the legitimate and
free exercise of religious belief." (*Lemon v. Kurtzman,*

403 U.S. 602, 622, 623, 91 S.Ct. 2105, 2116 [1971]. Emphasis added.)

The Church, in order to maintain her prophetic role, cannot be identified with any political party, with any national or international political institution, or with any economic system. The ministry of the Church may look favorably on the social-welfare program of the Democratic party and at the same time endorse the position of conservative Republicans on the abortion issue. The right wing of the Republican party was undoubtedly pleased with Pope John XXIII's position on the limited role of government, but it may be assumed that the need for socialization proclaimed in his *Mater et magistra* appealed more to left-wing Democrats.

The prime objective of the ministry of the Church is to help us to turn Christian principles into attitudes that motivate our conduct. These attitudes, which grow out of a faith that we are in God's hands and care, may be summarized by Martin Luther's two formulae: A Christian man is a perfectly free lord of all, subject to no one else; and, a Christian man is a perfectly free servant of all, subject to everyone, accountable to God. The activities of organized religion are not always motivated by these attitudes.

It is true that in this country religious values pervade the fabric of our national life, and that by raising our consciousness religion helps to identify the issues confronting our society. But it is also true that because of the freedom gained through religion, it is up to each individual to make up his own mind on public issues and on the merits of those who seek public

office. President Lincoln expressed his conviction in
the frailty of man and in the Almighty's "own
purposes," albeit in a different context, when he
wrote: "Men are not flattered by being shown that
there has been a difference of purpose between the
Almighty and them. To deny it, however, in this case,
is to deny that there is a God governing the world."
No man can claim the exclusive ability to grasp God's
purposes and to impose his own perspective on everyone else. The Old and New Testaments provide us
with the ideas of individual freedom, of individual
conscience, of tolerance based on a belief in the dignity of the individual, and of a moral authority transcen-
ding any office and any boundary.

1. The First Amendment

The premise that religion is a source of freedom is
in no way weakened by contemporary controversies
concerning the meaning of the First Amendment.
Yes, we may be puzzled by court decisions holding
that it is unconstitutional to permit students to join in
a nondenominational prayer or in a minute of silent
meditation, while the aid of God is invoked in legislative sessions. We will continue to witness debates on
the interpretation of the religious clauses of the First
Amendment—debates seeking to find answers to such
issues as whether these clauses permit federal tax
credits to parents who send their children to private
religious schools where religious creeds are taught;
whether the appointment of a U. S. ambassador to the
Vatican is unconstitutional since, in doing so, an
"official relationship" with the religion may be

established; whether the adoption of a proposed amendment permitting sanctioned group prayer would drastically alter the First Amendment, which commands the government to be strictly neutral in respect to religion; or whether posting the Ten Commandments in classrooms or offering grace before meals endangers the preservation of religious freedom guaranteed by the First Amendment to all Americans of all faiths.

It should be stressed that these debates caused by Court decisions, which appear to some to be intended to guarantee freedom from religion rather than freedom of religion, deal with the interpretation of religious clauses of the First Amendment and not with religion as a basis of our freedoms.

The desire for religious liberty which animated the founding of our Republic—and which produced the Declaration of Independence, the Constitution, and the Bill of Rights—did not imply an antagonistic attitude toward religion on the part of the Founding Fathers. Nor did the Founding Fathers mean to preclude the application of religious principles to public life. Although the First Amendment banned the establishment of a national religion—the dangers of which having been so amply demonstrated for centuries—and maintained governmental neutrality among various denominations and sects, it was never meant to hinder religious practice nor to prevent religion from playing a part in the life of the nation. It was expected that religious convictions would influence both the private lives of individuals and the collective life of society. Thus, in 1787, the promulgators of the Ordinance for the Government of the Northwest Ter-

ritory declared that "religion, morality, and knowledge being necessary to good government and the
happiness of mankind, schools and the means of education shall forever be encouraged."

The deeper purpose of the First Amendment was to
reserve from all official control the domains of intel-
lect and conscience—and conscience is often the echo
of religious faith. The question as to which belief in a
superhuman power is correct was placed beyond the
jurisdiction of any court or legislature. Thus the
privacy of personal religious experience and the liberty of all religious performance were assured. The First
Amendment was designed to protect freedom of conscience, not to abolish or diminish the roles of conscience and belief in society. As Chief Justice Hughes
stated in the *Macintosh* case: "But in the form of
conscience, duty to a moral power higher than the
state has always been maintained." (283 U.S. 605,
633, 51 S.Ct. 570, 578 [1930].)

The freedom of belief which the First Amendment
protects from State action includes both the right to
speak freely and the right to refrain from speaking.
These rights are essential components of a broader
concept of freedom of thought; there can be no freedom of thought unless ideas can be uttered. This is in
turn a part of a still broader conception of human freedom and dignity. Those who drafted the Bill of Rights
realized that the government they were creating was
not the source of the spirit of freedom they were try-
ing to protect. No mere legal document could be the
source of an individual's rights, duties, dignity, and inherent equality with other individuals—these come
from the Creator. Man was meant to be free and to

have the ability to stand unimpeded in the light of his Creator as he saw fit to discern that light. The "wall of separation" erected by the First Amendment to the Constitution of the United States was intended to keep government from intruding into that light—and not to keep that light from illuminating the fields of public discourse, as too many have tried to interpret that intent.

2. Unchangeable Values in a Changing Society

Because our courts, especially the Supreme Court, are not insensitive to the public mood, the changes that occur in our society have an impact on the courts' opinions. The judges often adapt their decisions to what Justice Holmes called "the felt necessities of the time." Although judges are generally supposed to adhere to continuity by affirming the findings of their predecessors, the way the courts have interpreted the meaning of religion as a source of our freedom has not been fixed and invariable, due to the constantly changing political forces to which the courts react.

In the nineteenth century our courts took the position that Christianity had entered and influenced, more or less, all our institutions, and it was "interwoven into the texture of our society." According to one court decision, "it is involved in our social nature, that even those among us who reject Christianity, cannot possibly get clear of its influence, or reject these sentiments, customs and principles which it has spread among the people, so that, like the air we breathe, they have become the common stock of the

whole country, and essential elements of its life."
(*Mohney v. Cook*, 26 Pa. 342 [1855].)

Until World War I Christianity—although not the
national religion—was the principal religion, a part of
common law, "the purest system of morality, the
firmest auxiliary, and only support of human laws."
(*Updegraph v. Commonwealth*, 11 Serg. 394, 406 [Pa.
1824].) After World War I, the doctrine of the State's
neutrality in its relations with groups of religious
believers and nonbelievers took the place of belief in
religion as a source of our God-given liberties. Strik-
ing down as unconstitutional a state law prohibiting
the teaching of the doctrine of evolution in public
schools, the Supreme Court stated: "Government
in our democracy, state and national, must be neutral
in matters of religious theory, doctrine, and practice.
It may not be hostile to any religion or to the advocacy
of no-religion; and it may not aid, foster, or promote
one religion or religious theory against another or
even against the militant opposite. The First Amendment mandates governmental neutrality between religion and religion, and between religion and nonreli-
gion." (*Esperson v. Arkansas*, 393 U.S. 97, 103-104,
89 S.Ct. 266, 270 [1968].)

Justice Black in a dissenting opinion once
said: "The First Amendment has lost much if the religious follower and the atheist are no longer to be
judicially regarded as entitled to equal justice under
law." (*Zorach v. Clauson*, 343 U.S. 306, 320, 72 S.Ct.
679, 687 [1952].) In a further attempt to sever our
freedom from its religious roots, the Court declared
that neither the federal nor a state government can
"constitutionally pass laws or impose requirements

which aid all religions as against non-believers, and neither can aid those religions based on belief in the existence of God as against those religions founded on different beliefs." (*Torcaso v. Watkins*, 367 U.S. 488, 495, 81 S.Ct. 1680, 1683-4 [1961].) Among the "religions" in this country which do not teach what might generally be considered a belief in the existence of God are Buddhism, Taoism, Ethical Culture, Secular Humanism, and others.

Faced with the erection of wider and higher walls of separation of our freedom from its source, a religion "based on belief in the existence of God," we should be reminded of the reaction of President Lincoln to the Supreme Court's decision in *Dred Scott vs. Sandford*. (19 How. 393-638, 15 L. Ed. 691 [1857].) In this case the Supreme Court declared in effect that under the Constitution Negro slaves were property and that Congress could not, under the Fifth Amendment, constitutionally deprive slaveholders of this property. The authority of the Congress to restrict the spread of slavery was thus preempted. Reacting to this decision President Lincoln argued, "If this important decision had been made by the unanimous concurrence of the judges . . . [or if] it had been before the court more than once, and there been affirmed and reaffirmed through a course of years, it might then be, perhaps would be, factious, nay, even revolutionary, to not acquiesce in it as a precedent. But when as it is true we find it wanting all these claims to the public confidence, it is not factious, it is even not disrespectful, to treat it as not having yet quite established a settled doctrine for the country."

The establishment of a secular basis for freedom is

far from becoming a "settled doctrine for the country." The Court itself has demonstrated that our society must no longer expect that all principles must adhere to precedent, since the judicial process can be adapted to varying conditions. The Supreme Court of the United States has explicitly asserted its right to overrule a prior constitutional decision when it realizes that the prior principle is wrong. (*Smith v. Allwright*, 321 U.S. 649, 64 S.Ct. 757 [1944].)

There are signs of a shift in the Supreme Court's position toward official acknowledgement of the role of religion in American life. The Justices of the Supreme Court, Justice Frankfurter once wrote, "breathe life, feeble or strong, into the inert pages of the Constitution and of statute books." In the "breathing" of the Court under Chief Justice Warren Burger a trend may be discerned toward the accommodation of religion to public life, as long as there is no "real danger of establishment of a state church." Such an accommodation recognizes that political issues also have moral and ethical dimensions which have always been and will continue to remain a part of the concerns of religion.

It is not always easy to draw a line of distinction between a private, individual morality and public morality, since a member of a church is also a member of society and a citizen of the state. In 1971 the Supreme Court recognized that under the Establishment and Free Exercise Clauses of the First Amendment "total separation [of Church and State] is not possible in an absolute sense. Some relationship between government and religious organizations is inevitable." In the absence of precisely stated constitutional prohibitions by the Religion Clauses of the First Amendment, in order to determine whether a challenged law or conduct establishes a religion or religious faith, the Court found it useful to apply a three-pronged test reflecting the cumulative criteria developed by the Court over the years. In scrutinizing a challenged conduct or legislation, the three-pronged analysis serves as a guidance to inquire: First, whether the challenged law or conduct has a secular purpose; secondly, whether its principal or primary effect is to advance or inhibit religion; and finally, whether it creates an excessive entanglement with religion. (*Lemon v. Kurtzman*, 403 U.S. 602, 91 S.Ct. 2105 [1971].)

Ten years later, in 1981, the Supreme Court stated that a state university's institutional mission to provide a secular education to its students does not exempt from constitutional scrutiny its denial of the use of university facilities to a registered student religious group. (*Widmar v. Vincent*, 454 U.S. 263, 102 S.Ct. 269 [1981].) The object of scrutiny was the action brought by members of a registered student group named Cornerstone at the University of Missouri at Kansas City (a state university), challenging the university's policy of excluding religious groups from the university's open forum policy, whereby university facilities were generally available for activities of other registered student groups. (The university officially recognized over 100 groups.) This discriminating exclusion was based on the university's regulation prohibiting the use of university buildings or grounds "for purposes of religious worship or religious teaching."

The student members of Cornerstone alleged that

this regulation discriminated against religious activity and discussion and violated their rights to free exercise of religion, equal protection, and freedom of speech under the First and Fourteenth Amendments. The Supreme Court held that: (1) having created a forum generally open for use by student groups, the university was required to justify its discriminations and exclusions under applicable constitutional norms; and (2) the challenged exclusionary policy violated the fundamental principle that State regulation of speech should be content-neutral. The decision declared that the First Amendment rights of speech and association extend to the campuses of state universities. Religious worship and discussion are forms of speech and association protected by the First Amendment.

In March 1984, in *Lynch v. Donnelly*, the Supreme Court held that the Constitution does not require complete separation of Church and State; it affirmatively mandates accommodation, not merely tolerance, and forbids hostility toward any. "There is an unbroken history," wrote the Court, "of official acknowledgement by all three branches of government of the role of religion in American life from at least 1789." The concept of a "wall" of separation between Church and State is, therefore, "a useful metaphor" but "is not a wholly accurate description of the practical aspects of the relationship that in fact exists. . . ." (104 S.Ct. 1355, 1359 [1984].)

In this case the Court held that the city of Pawtucket, Rhode Island, could continue to display a crèche, or Nativity scene, which was municipally owned and that for over forty years had been a part of an annual display in a park owned by a nonprofit organization and located in the heart of the city's shopping center. Some Pawtucket residents and individual members of the Rhode Island affiliate of the American Civil Liberties Union brought the court action challenging the city's inclusion of the crèche in the annual display. Wrote Chief Justice Burger: "We can assume, *arguendo*, that the display advances religion in a sense; but our precedents plainly contemplate that on occasion some advancement of religion will result from governmental action." Additionally, the Court has made in abundantly clear that "not every law that confers an 'indirect,' 'remote,' or 'incidental' benefit upon religion is, for that reason alone, constitutionally invalid." (*Ibid.*, 1364.)

The Court concluded that to forbid the use of the crèche, which is a passive symbol, "would be a stilted over-reaction contrary to *our history and to our holdings.*" If the presence of the crèche in this display, the Court pointed out, violates the Establishment Clause, "a host of other forms of taking official note of Christmas, and of our *religious heritage*, are equally offensive to the Constitution." (*Ibid.*, 1365, emphasis added.)

The shift in the Supreme Court's position toward official acknowledgement of the role of religion in American life has not undermined the separation of Church and State nor lowered the constitutional barriers that prohibit the state and federal governments from sponsoring, promoting, participating in, or financially supporting religious activities. In 1985 the Court affirmed Church-State separation by voiding an Alabama statute allowing moments of silence in public schools, by scrapping well-established "share-time" programs in which public-school teachers taught part-time special classes in parochial schools, and by declaring unconstitutional a Connecticut law requiring employers to give employees days off on their Sabbaths (a privilege that might conflict with the preferences of others). The Establishment Clause continues to insist on the neutrality of federal and state governments in religious matters.

The metaphor of a "wall" (deriving from Thomas Jefferson's reply to an address by a committee of the Danbury Baptist Association, January 1, 1802) serves as a reminder that the object of the First Amendment was to prevent the formation of any national ecclesiastical establishment that had the exclusive patronage of the national government. But the fact remains that we do not live in a vacuum, and an attempt to enforce a regime of total separation of Church and State, a separation never intended by the First Amendment and uniformly rejected by the Court, will only undermine our national traditions, our religious heritage, and the ultimate constitutional objective of the First Amendment.

This fact remains true in the technological age with its message that change is ever imminent. We must be prepared to accept changes, but ethical ideas, including the concept of freedom based on religion, will remain the most important formative influences on human conduct. Their influences should disperse the fears that technology will dehumanize our environment. Computers may improve our life style, but the speed and volume of instant information they provide do not exhaust the purpose of life. Greatness is achieved not by high standards of living but by high standards of life based on moral values.

Traditional moral values will continue to exercise their formative influences on human behavior. The identical meaning of such terms as "religious belief," "the voice of conscience," and "moral obligation," which often are used interchangeably, points to the close relationship between morality and religion. "For what is religion," asked the court, "but morality with a sanction drawn from a future state of rewards and punishment?" (*McAllister v. Marshall*, 6 Bin. 338, 6 Am. Dec. 458 [Pa. 1814].) On August 10, 1787, in a letter to his nephew Peter Carr, Jefferson wrote:

> "Man was destined for society. His morality therefore was to be formed to this object. He was endowed with a sense of right and wrong merely relative to this. This sense is as much a part of his nature as the sense of hearing, seeing, feeling; it is the true foundation of morality, and not the . . . truth, &c., as fanciful writers have imagined. The moral sense, or conscience, is as much a part of man as his leg or arm. It is given to all human beings in a stronger or weaker degree, as force may be strengthened by exercise, as may any particular limb of the body. This sense is submitted indeed in some degree to the guidance of reason; but it is a small stock which is required for this: even a less one than what we call Common sense. State a moral case to a ploughman and a

professor. The former will decide it as well, and often better than the latter, because he has not been led astray by artificial rules."

Some moral traditions undoubtedly have evolved through adaptation to the ends of life and are products of age-long experience. Some moral traditions such as customs, which we now call manners, are closely linked to the evolution of human institutions. It is also undeniable, however, that traditional moral values such as freedom, justice, prudence, love, fidelity, honesty, and other similar values recognized by Plato, Aristotle, and Moses, and by Christ's teaching as the virtues that develop the whole man, remain unchangeable. They are the same today in the age of technology as they were yesterday. (For a more detailed study of morality and religion, see Chapter 1 of my book *The Foundations of a Free Society*, The University of Texas at Dallas, 1983, pp. 27-65.)

B. Education

When we list education as a source of freedom we should not overlook the fact that most of early educational efforts in the United States were motivated by religion. Out of the 246 colleges founded prior to 1846, only 17 were started by the states; the rest were founded by religious groups. The textbooks in use, like the *New England Primer* which appeared in Boston around 1690 or the reader compiled by William Holmes McGuffey, were religious in nature with a strong emphasis on moral values and on building a society of free men.

The price of freedom, as we mentioned, is the citizen's eternal vigilance against the dangers of freeing his government to become its own god. "Every government," wrote Jefferson, "degenerates when trusted to the rulers of the people alone. The people themselves therefore are its only safe depositories. And to render even them safe their minds must be improved to a certain degree." In exercising this vigilance, education has the greatest relevance. Jefferson's Bill for the More General Diffusion of Knowledge, along with two other bills—one for the creation of a public library and another for a reform of the College of William and Mary—were aimed at the establishment of an educational system designed to ensure the educational opportunities that are indispensable in preserving a free society.

In Jefferson's pyramidal plan of educational opportunities for all free men, elementary schools offering reading, writing, and common arithmetic would be available to all children, regardless of ability to pay; higher levels of the system offering Greek, Latin, geography, "higher branches of arithmetic," and sciences would be open to selected "geniuses" for their outstanding performance in the lower branches of learning. The close relation between democracy and education is expressed in Jefferson's proposal prescribing the selection "of the youth of genius among the classes of the poor." By such a selection, he hoped "to avail the state of those talents which nature has sown as liberally among the poor as the rich, but which perish without use, if not sought for and cultivated."

1. Crusade against Ignorance

The main purpose of diffusing knowledge through the mass of the people, asserted Jefferson, is to render "the people safe, as they are the ultimate guardians of their own liberty." He saw no reason why 160,000 electors in the island of Great Britain should give the law to four million in the states of America where every individual is equal to every elector "in virtue, in understanding, and in bodily strength." He accused the government of Great Britain of corruption because "but one man in ten has the right to vote for members of parliament," since the right of voting was confined to a few of the wealthier of the people. To avoid such corruption Jefferson called for a crusade against ignorance to "establish and improve the law for educating the common people."

This crusade remains pertinent to our republican form of government, where all citizens have a right to vote. Because of the enormous heterogeneity of our nation with respect to political views, religious beliefs, and moral codes, the voter is subjected to a barrage of assertions, arguments, and intentional or unintentional attempts at persuasion. When issues are simplified by labeling them with terms that have a favorable or unfavorable connotation—such as "liberal" and "conservative," "Christian" and "communist"—only an educated citizen can discern the truth and prevent emotions from triumphing over reason and good judgment. Only an educated citizen will reject the labels affixed by some upon those whose ideas they oppose.

The Nineteenth Amendment to the Constitution

states, "The right of citizens of the United States to vote shall not be abridged by the United States or by any state on account of sex." With federalized citizenship, the right to vote is a federal constitutional right. The importance of this voting-right to democracy and to a free, responsible, and statesmanlike government was described in a colorful way by the Court of Appeals of Kentucky:

> "The woman with the sun bonnet and the checkered apron who trudges off of the mountain side in Leslie County and walks down the creek a mile to cast her vote—she is an American queen in calico, but her only pay for voting is the satisfaction of knowing that Columbia, by God's help and hers, shall continue as the gem of the mighty ocean. Let no man cease to thank his God as he looks in at the open door of his voting place, as he realizes that here his quantity, though cast in overalls, is exactly the same as the quantity of the President of the United States. There is a satisfaction and privilege in voting in a free country that cannot be measured in dollars and cents." (*Illinois Central Railroad Company v. Commonwealth*, 204 S. W. 2d 973, 334 U.S. 843 [1947].)

Civic zeal is inseparable from the exercise of political rights. Describing the public spirit of the United States, Tocqueville wrote, "I maintain that the most powerful, and perhaps the only, means which we still possess of interesting men in the welfare of their country is to make them partakers in the government."

This "satisfaction and privilege in voting" is endangered by suppression of public discussion. Free political discussion is a fundamental principle of our constitutional system, since its end is a government responsive to the will of the people. Such free political discussions take place within the major political parties in the United States. The party's platform is determined through the exchange of often conflicting ideas, and the party membership as a rule is held together by acceptance of general principles, by the cohesive influence of leaders, and sometimes by the power of patronage. A citizen who votes for a party ticket does not have to apply for membership or to pay dues. He and the elected representatives of his choice remain free to speak and act as their consciences demand. None of them is committed to take orders from the party's leaders or to execute their plans. One may disagree with the party's policies, leave it, and return as he wills.

Referring to the major parties in Great Britain, whose system in some ways resembles our own, Lord Balfour wrote: "Our alternating Cabinets, though belonging to different parties, have never differed about the foundation of society, and it is evident that our whole political machinery presupposes a people so fundamentally at one that they can afford to bicker; and so sure of their own moderation that they are not dangerously disturbed by the never-ending din of political conflict. May it always be so."

Individual nonconformity and moderation, so essential to democracy, are not tolerated by totalitarian countries. In Germany, when Nazi party members married, they received a copy of *Mein Kampf* as a wedding gift from the party to remind the newlyweds of their obligation to produce future Nazis. From preschool onwards, children were taught that the State is more important than the individual, and that the individual must be ready to sacrifice himself for fatherland and Hitler. Party members pledged strict discipline, absolute obedience, and allegiance to one man, an absolute dictator.

The members of the Communist party in Soviet Russia and elsewhere in the world are fully committed to the policies of the party. They carry on a continual purge and ruthlessly remove from office anyone who shows the slightest sign of deviating from the party line. Schools, from the elementary grades to the most advanced studies, are transformed into agencies for State training closely resembling military agencies for training and mobilization. Academic freedom, in the sense that it is practiced in the Western world, is unknown. The motivating force of Soviet education is to produce a commodity which will best advance their system and establish the USSR as a dominant power in the world.

2. Intolerance of Freedom

To avoid the tragedy of that sort of dictatorship, Jefferson prescribed for the ordinary man chiefly historical studies. He confirmed the truth of the old maxim, *Historia mater studiorum est* ("History is the mother of all studies"), when he wrote, "History by apprising them [men] of the past will enable them to judge of the future: it will avail them of the experience of other times and other nations; it will qualify them as judges of the actions and designs of men; it will enable them to know ambition under every disguise it may assume; and knowing it, to defeat its views."

A student of history will find that the Free World is constantly exposed to the demands and the hypocritical resolutions of Communist parties against alleged infringement of their freedom. Their concept of freedom is to have for themselves the free opportunity to impose upon our society an organization built on principles fundamentally opposed to those offered by our Constitution. Their concept of freedom means intolerance of the freedom we enjoy and the coerced adoption by each individual of the ideas of the ruling group in power.

History shows us that in not one of the countries under communist dictatorship was the government chosen by free election. The communists seize power and install a government by and for a minority by means of deception, terrorism, assassination of their opponents, and *coup d'état*. The rule of a minority, based on totalitarian disciplines and techniques, brought the Bolsheviks to power in Soviet Russia. Following the overthrow of the Romanov dynasty, the moderate socialists defeated the Bolsheviks in the election of November 1917 by a majority of almost two to one. When the assembly convened in the middle of January 1918, the Bolsheviks dissolved it by force at its first session. Since 1917 the goal of the Communist party has remained the same—to seize the reins of government by a minority who incite prejudices, discontent, and civil disorder, spread untruths, and discredit the existing governments of the day.

History enables us "to know ambition under every disguise it may assume." The Communist parties throughout the world serve as "fifth columns" for the Kremlin's imperialistic ambitions. Before World War II, American communists in synchrony with Soviet policy, were anti-German, "anti-fascists." Upon Stalin's conclusion of a nonaggression pact with Hitler, the communists in the United States and elsewhere denounced our aid to the victims of German aggression. When Germany attacked Russia, the communists echoed Russia's demand for immediate aid to Russia and for opening a second front in order to relieve the military pressure on the Soviet Union. This satrap party—one controlled by a foreign government—still echoes its master and tries to conceal its intolerance of freedom by proclaiming grievances which throw blame upon those who engage in resistance to or reprisals against attempts to overturn our system of free self-government. Judge Medina, in addressing this issue, asked: "What is freedom anyway? I thought I knew something about it, until I went through the trial of the Communist leaders back in 1949, and was subjected to a barrage of propaganda concerning the so-called freedom of the Russian variety." ("The Bill of Rights—Our Heritage," *Vital Speeches of the Day*, Vol. XXIV, No. 8, p. 254.)

In our zeal to defend ourselves against the onslaught of the communists, we must also protect ourselves against the extremists ready to label every "liberal-minded" person as a "communist." Justice Jackson expressed the need for such protection in his opinion concurring and dissenting (each in part) with the Court's decision concerning the constitutionality of the statute requiring labor union officers to forswear their membership in the Communist party. After giving a profound analysis of the goals of the Communist party, he concluded that "we are faced with a lawless and ruthless effort to infiltrate and disintegrate our society," and that, therefore, the Congress has the power to require labor union officers to disclose their membership in or affiliation with the Communist party. (*American Communications Association v. Douds*, 339 U.S. 382, 435, 70 S.Ct. 674, 702 [1950].)

Referring to other "fanatics and extremists," Justice Jackson also warned: "But we must not forget that in our country are evangelists and zealots of many different political, economic, and religious persuasions whose fanatical conviction is that all thought is divinely classified into two kinds—that which is their own and that which is false and dangerous." We must be alert to the danger of adopting the rules, techniques, and terroristic methods of totalitarian parties that must lead to the erosion of our much treasured heritage of freedom. If we allow ourselves to indulge in actions that are intolerant of the freedom of others, we will lose our own freedoms.

3. Classical and Scientific Studies

While Jefferson recommended for the "ordinary" man a curriculum of chiefly historical studies, for the intellectual elite he prescribed classical and scientific studies. Jefferson's wide horizon and vision put him in advance of his time. For the last two centuries scholars and authorities in education saw in the humanities the foundation of civilization. The importance of the humanities in enlarging and illuminating one's life cannot be denied. At least until recently, the humanities had a central place in the curricula of the colleges, universities, lycées, and gymnasia of Western Europe.

In the United States, on the other hand, the National Endowment for the Humanities in its 1984 study titled *To Reclaim a Legacy: A Report on the Humanities in Higher Education* asserted that the reason for the decline of undergraduate learning is the low estate of the humanities. To remedy this decline, the report prescribed the restoration of the humanities as the centerpiece of an undergraduate curriculum which should include the history of Western civilization; "a careful reading" of masterworks of European, English, and American literature; "demonstrable proficiency in a foreign language"; and a study of the "most significant ideas and debates in the history of philosophy." The question, however, remains whether a study of the humanities that embraces the disciplines of philosophy, art, literature, history, and languages builds a commitment to a democratic free society.

This question is also pertinent to the study of the sciences. In our age of technology, the sciences have brought changes of new dimensions along the whole front of our economy, have revolutionized our medical research, and have brought about sweeping changes in the social structure of our society. To science and research we owe great improvements in our health, the greatest industrial productivity known to man, and a knowledge of the laws which govern the universe. But science has also given us weapons that can destroy mankind. The question remains whether the impact of technological changes will, on the whole, be for good or for evil, and whether these changes will undergird our commitment to a free self-governing society.

History has not substantiated Jefferson's faith that the study of the humanities and of the sciences will safeguard free societies from corruption and despotism. In my book *The Foundations of a Free Society* (p. 170), I pointed out that the European educational system, with its sharp division between academic and vocational studies and its concentrated efforts in humanities and sciences, did not pass the crucial test of citizenship. The universities especially failed in their task of acting as a conscience for society. Those scholars who boasted of their devotion to the causes of truth and culture were silent when intellectual truth and moral freedom were suppressed by the corrupted Nazi regime. Reinhold Niebuhr, referring to the capitulation of German universities, wrote, "The culture of the university sought universal truth through the genius of the wise man; and forgot that the wise man is also a sinner, whose interest, passion and cowardice may corrupt the truth!" (*Beyond Tragedy*, Charles Scribner & Sons, 1937, p. 284.)

Jefferson called freedom "the firstborn daughter of science," but prophetically he also warned that it could "produce the bitter fruits of tyranny and rapine." The Soviet Union is investing more of its resources in scientific research than any other nation on earth, predominantly in the field of war. The trust in science as mankind's Messiah may put blinders on us, blinders that keep us from seeing the true role of learning in our determination to have a free self-governing society. Curricula concentrating on the humanities or on science only for their own sakes cannot achieve the goal of strengthening our faith in a free society, since knowledge without moral direction becomes aimless and meaningless. Freedom is not something that can be spoon-fed. It is a state of being, and it cannot flourish when independent thought and an understanding of the environment surrounding us are neglected. Respect for certain basic political values, such as freedom, is instilled not by abstract reasoning or by the accumulation of facts. It must be grounded in the true values of life, in the hopes and aspirations that govern our nation's destiny and its role in this world of ours.

The more the mind understands the forces and the order of a free society, the more it is able to analyze, to compare, and to distinguish, and through the process of critical inquiry to recognize the meaning of the democratic system, of our political heritage, and of the importance of moral values in guiding our actions. There is a definite relevance between a society's freedom and its educational system. Understanding leads to right action. Education's unique responsibility is to bring this understanding to nearly one third of the population of the United States; it involves the teacher, the student, and the administrator alike.

C. Law

The only limitation on individual freedom should be the freedom of others. While we strive for the greatest possible individual liberty from governmental authority, an individual's actions cannot adversely affect other members of society. Total freedom is the law of the jungle where only the fiercest, the strongest, and the most ruthless survive. Freedom is not absolute and is justified only if in pursuing our own interest we do not deprive others of theirs or impede their efforts to obtain it. Freedom does not relieve the individual from his duty to contribute to the general interest of the society in which he lives or from his duty not to injure others by his conduct. In a free society the rule of law curbs human aggression and directs it into constructive channels, thus becoming one of the pillars of civilization.

In the early history of mankind, religious and secular laws were inextricably mixed. Laws were accepted as divine revelation. The Decalogue combines the fundamental articles of Israel's laws with religious and ethical obligations. It combines duties to God (1-4) with duties to man (5-10). In the "Preamble,"—"I am Jehovah thy God, who brought thee out of the land of Egypt, out of the house of bondage" (Exodus 20:2)—God revealed himself as Jehovah, the Deliverer and the Redeemer. Since He redeemed Israel from slavery and serfdom, and all the land was owned by God, He had the right to give them commandments that should constitute the very heart and the very essence of human behavior. The laws binding the Jews of the Old Testament were, therefore, interwoven with their religious beliefs.

The law commanding "Honor thy father and thy mother" was a recognition of a child's obligation to legitimate authority—the parents—who, according to the ancients, stood to their children in place of God Himself. Filial disobedience was not only a violation of law but also a sin against God, violation of a religious duty. The remaining commandments (6-10) include laws that give a new conception of the sacredness of life, of the sanctity of marriage, of the right to property, of the virtue of veracity directed against perjury and untruthfulness when on the witness stand, and of the illicit longing that is often the root of illicit conduct. The Decalogue made law a part of religion.

The pre-Hebrew code of law called the Code of Hammurabi has many similarities with Jewish law. Some scholars maintain the latter was patterned after such a Babylonian code. The Koran, which accepted the teaching of the Jewish prophets (and considered Jesus one of them), was influenced by the Old Testament's moral and ethical precepts. Jesus, engaged in teaching rather than in lawmaking, upheld Jewish law by declaring in his Sermon on the Mount: "Think not that I am come to destroy the law, or the prophets; I am not come to destroy, but to fulfill" (Matthew 5:17). The Mosaic law found in the Pentateuch and the New Testament does not make a distinction between religious and civil ordinances; they formed one legal system. Jesus did not change the law but directed its interpretation into new channels of love, forgiveness, humility, and generosity.

In ancient Greece, law was also based on a religious foundation, albeit a polytheistic one. In the field of law the jurisdiction was divided among the numerous gods. Lawmakers Draco and Solon as well as the Greek philosophers Socrates, Plato, and Aristotle believed that the universe was governed by a sort of higher law. According to Plato, "No law or ordinance has the right to sovereignty over true knowledge." To the Greeks, God was all-pervading in the world, and the universe, or cosmopolis, was governed by Logos—the infinite total of all ideas discoverable by reasoning. Logos was theoretically a form of natural law. In Rome the natural law theory found eloquent and brilliant supporters and among them its most prominent exponent was Cicero, who saw in law "nothing else but high reason, calling us imperiously to our duty, and prohibiting any violation of it." As a diligent student of Greek philosophy he tried to discover the rules governing the world, and in natural law he saw the reflection of divine and eternal law revealed to man through reason.

The concept of natural law assumes that there is a certain standard of morality upon which all legal systems rest. Because of the intimate connection between natural law and morality, law may be considered as one of the sources of freedom only after defining the standards of morality upon which the law was founded. These standards, however, vary, and with them the meaning of natural law varies as well.

Thomas Aquinas's theory of natural law, for instance, rests on a theological base. For Saint Thomas, God is the ultimate sanction for law and morality, and natural law was ordained by Him and revealed to man through reason. Aquinas saw a close relationship between "eternal law," or "natural law," and human law, and expected the latter to be "a dictate of reason" and to serve the "well-being" of the citizens. For this reason, "tyrannical law, not being according to reason, is not law at all in the true and strict sense but is a perversion of law." On a similar note, Saint Augustine concludes that "that which is not just seems to be no law at all."

Hobbes also defines natural law as a "dictate of right reason," but for him "right reason" is not a reflection of eternal truth but "the act of reasoning by which man uses all means to insure his survival." The rules of human behavior deduced from human nature are, according to Hobbes, the laws of nature. Hobbes saw in natural law and civil law, "not different kinds of Law but different parts of Law, whereof one part being written is called Civil, the other unwritten, Natural." While for the church fathers natural law was absolute and invariable in its content, for Hobbes natural law—since it derives from human nature—changes with changing modes of human behavior. Human horizons, he asserts, are extended because "time and identity produce every day new knowledge."

Hobbes's principal distinction between natural law and civil law—that natural law is unwritten whereas civil law is promulgated in a written form—did not appeal to John Locke, who believed that men were endowed with certain natural rights even before there was a State. The state of nature, he argued, "has a law of nature to govern it, which obliges everyone: and reason, which is that law, teaches all mankind, who will consult it, that all being equal and independent, no one ought to harm another in his life, health, liberty, or possessions."

Locke greatly influenced Jefferson, and this influence can be traced through the Declaration of Independence. In its very first sentence we find reference to "the Laws of Nature and of Nature's God," which impelled the people of this nation to separate from England. The American Revolution that began in 1776 was the first historical event of the modern era that derived from a belief in individual rights. The Declaration asserted that human rights could not be abrogated by any human enactment because all men "are endowed by their Creator with certain unalienable Rights." As examples of "unalienable Rights," the Declaration gave "Life, Liberty, and the pursuit of Happiness."

Locke also influenced Jefferson with his theory of social contract. To secure these "unalienable Rights," "Governments are instituted among Men, deriving their just powers from the consent of the governed." Individuals do not owe their rights to the government; the government owes its whole legitimacy to the individuals, who have the right to abolish a government that has become "destructive," and to "institute new Government." These laws deriving from the "Laws of Nature and of Nature's God" are a source of our freedom. The right to be free is one of our "unalienable Rights," remaining constant anytime and anywhere. Our rights also carry with them the force of moral obligation.

1. The Power of Rediscovery: Natural Law

Jefferson insisted that amendments to the Constitution be made to include a declaration of rights. This Bill of Rights, the first ten amendments to the Constitution, was adopted to prevent "misconstruction" or "abuse of powers" and to extend "the ground of public confidence in the government." Jefferson was a profound political philosopher but not always a consistent thinker. In his writing we may find rhetorical excesses. His rhetoric, for instance, carried him to the point of advocating the need for revolution every twenty years to nourish the tree of liberty; to preferring newspapers without a government over a government without newspapers; and to declaring that every constitution, and every law, "naturally expires at the end of nineteen years." Such prophecies, if fulfilled, would have destroyed the American republic and the democracy rooted in our Constitution.

Moral values embodied in natural law are universal and invariable but human ideas are no more infallible than men. Generally accepted opinions in one age may be rejected by future ages as false and absurd. Rules of conduct are affected by a variety of causes that bring changes not always supported by reason. Among them are rules emanating from class interests, from sympathies and aversions that are not always explainable by reason but often constitute a significant force in molding the moralities and laws governing a society.

The most laudable changes are those resulting from the search for truth. Such a search is often accompanied by persecution; we have only to mention the

condemnation of Socrates or the suffering of the early
Christians. The dictum that truth must triumph only
after an ordeal of persecution does not have much appeal since it calls for sacrifices. What is appealing in
the search for truth may be extinguished temporarily
by oppressing forces, but it will continue to reappear
in more favorable circumstances until it prevails over
attempts to suppress it. Our jurisprudence pertinent
to race problems is an outstanding example of this
"power of rediscovery" that justice founded in equity,
in honesty, and in fairness possesses.

Justice   Holmes   wrote: "Every   important
principle which is developed by litigation is in fact at
bottom the result of more or less definitely understood views of public policy; most generally, to be
sure, under our practices and traditions, the unconscious result of intuitive preferences and inarticulate
convictions, but none the less traceable to views of
public policy in the last analysis." (*The Common Law*,
Little, Brown, and Company, 1881, p. 35.) How
changeable these "public policies" can be is shown by
the opinions concerning race relations delivered by
the Supreme Court in the cases of *Scott v. Sandford*
in 1856, *Plessy v. Ferguson* in 1896, and *Brown v.
Board of Education of Topeka* in 1954.

In 1856 in the case of *Dred Scott v. John F. A.
Sandford* (previously cited), the Court declared that
the Missouri Compromise—which prohibited slavery
in the portion of Louisiana Purchase Territory north of
Missouri—was unconstitutional because it deprived
slave owners of their property without due process.
The matter at issue before the Court was whether
descendants of Africans who were brought to this

country and sold as slaves could become citizens of the United States when they were emancipated or born of parents who had become free before their births. (The plaintiff was a citizen of the state of Missouri.) Chief Justice Taney delivered the opinion of the Court that they could not become citizens of the United States. They were not included, stated the Court, "and were not intended to be included under the word 'citizens' in the Constitution, and can, therefore, claim none of the rights and privileges which that instrument provides for and secures to citizens of the United States."

The Court took the position that the Negroes, descendants of slaves, were a subordinate and "inferior class" of beings, and, whether emancipated or not, they did not have the rights and privileges of "the people of the United States" who held the power and conducted the government through their representatives. In the opinion of the Court, the men who framed the Declaration of Independence intended to perpetuate the "impassable barrier" between the white race and the whole enslaved African race (including mulattoes). The Court further concluded that Negroes are "altogether unfit to associate with the white race, either in social or political relations; and so far inferior, that they had no rights which the white man was bound to respect. . . ." (*Ibid.*, at 407.) No state law, therefore, passed after the Constitution was adopted, could give any right of United States citizenship to members of the "enslaved African race," although individual states were (as the opinion recognized) competent to confer on such persons state citizenship effective within the state's boundaries.

The Court's position was challenged by many prominent lawyers in pursuit of justice. Mr. Justice Curtis in a dissenting opinion stressed that the Constitution of the United States was established by the people of the United States through the action, in each state, of those persons who were qualified by its laws to act thereon, in behalf of themselves and all other citizens of the state. In some of the states, members of the African race were among those qualified by law to act on this subject. "It would be strange," wrote Mr. Justice Curtis, if we were to find in the Constitution "anything which deprived of their citizenship any part of the people of the United States who were among those by whom it was established."

As to the intention of the men who framed the Declaration of Independence, Justice Curtis believed that—because of their strong belief in the universal truth that all men are created equal and are endowed by their Creator with unalienable rights, and because of the individual opinions they expressed and actions they undertook—"it would be not just to them, nor true in itself, to allege that they intended to say that the Creator of all men had endowed the white race exclusively, with the great natural rights which the Declaration of Independence asserts."

In 1896 in the case of *Plessy v. Ferguson*, nearly a half century later and thirty years after the adoption of the constitutional amendments which had clarified the status of Black Americans as citizens, the United States Supreme Court reviewed the constitutionality of an act of the General Assembly of the State of Louisiana, passed in 1890, providing for separate railway carriages for the white and colored races. The constitutionality of this act was challenged on the grounds that it conflicted with both the Thirteenth Amendment to the Constitution, which abolished slavery, and the Fourteenth Amendment, which prohibits certain restrictive legislation on the part of the states. The Court, by applying the doctrine of "separate but equal," found that the Statute of Louisiana, Acts of 1890, Number 111, was not in conflict with the Constitution.

In *Plessy v. Ferguson*, the Court took the position that the object of the Fourteenth Amendment was to enforce equality of the two races before the law, but it was not intended to abolish distinctions based upon color or to enforce social, as distinguished from political equality, "or commingling of the two races upon terms unsatisfactory to either." Ignoring totally the reality of the sentiments caused by separation of the two races, the Court remarked: "We consider the underlying fallacy of the plaintiff's argument to consist in the assumption that the enforced separation of the two races stamps the colored race with a badge of inferiority. If this is so, it is not by reason of anything found in the act, but solely because the colored race chooses to put that construction upon it." (163 U.S. 536, 551, 16 S.Ct. 1138, 1143 [1896].)

The Court held that it was within the competency of the state legislatures in exercising their police powers to establish laws permitting separation of the two races in places where they were liable to be brought into contact. The common instances of the valid exercise of such legislative power were the establishment of separate schools for white and for colored children and the separation of the two races in theaters and railway carriages. Such separation, according to the Court, did not stamp the colored race with the stigma of inferiority.

In the *Plessy* case, the Court suggested that legislatures were powerless to eradicate racial instincts or to abolish distinctions based upon physical differences, "and the attempt to do so can only result in accentuating the difficulties" in race relations. According to the "separate but equal" doctrine applied by the Court, the government has performed all of the functions respecting "social advantages with which it is endowed" when it has secured to each of its citizens equal rights before the law and equal opportunities for improvement and progress. "If," stated the Court, "the civil and political rights of both races be equal one cannot be inferior to the other civilly or politically. If one race is inferior to the other socially, the Constitution of the United States cannot put them upon the same plane."

The famous dissent by Mr. Justice Harlan in this case was the voice of natural law, or natural justice, protesting the injustice of the "separate but equal" doctrine and its inconsistency with the Constitution of the United States. The Thirteenth and the Fourteenth amendments, he claimed, decreed universal civil freedom in this country and removed the race line from our governmental system. "The white race deems itself to be the dominant race in this country," he wrote,

"But in view of the Constitution, in the eye of the law, there is in this country no superior, dominant class of citizens. There is no caste here. Our Constitution is color-blind, and neither knows nor tolerates classes. In respect of civil rights, all citizens are equal before the law. The humblest is the peer of the most powerful. The law regards man as man, and takes no account of his surroundings or of his color when the civil rights as guaranteed by the supreme law of the land are involved." (*Ibid.*, at 558.)

For over half a century, the American courts labored with the doctrine of "separate but equal" without an attempt to reexamine the doctrine in its application to public education. Only in 1954 in the case of *Brown v. Board of Education of Topeka* did the Supreme Court decide to reconsider the question whether segregation of children in public schools solely on the basis of race—even though the physical facilities and other "tangible" factors might be equal—deprived the children of the minority group of equal educational opportunities. The Court stated that it did and concluded that in the field of public education the doctrine of "separate but equal" had no place. Mr. Justice Warren, who delivered the opinion of the Court, wrote:

"To separate school children from others of similar age and qualifications solely because of their race generates a feeling of inferiority as to their status in the community that may affect their hearts and minds in a way unlikely ever to be undone. . . .
"Whatever may have been the extent of psychological knowledge at the time of *Plessy v. Ferguson*, this finding is amply supported by mod-

ern authority. Any language in *Plessy v. Ferguson* contrary to this finding is rejected." (347 U.S. 483, 494-495, 74 S.Ct. 686, 691-692 [1954].)

Decisions of the courts are reversed as our perceptions of justice and natural rights develop. In our search for a truth that may be discovered by reason and revelation, we are far from claiming that the *Brown* decision or subsequent civil rights legislation offers an absolute solution to the problem of race relations in the United States. The remedies offered for "historic discrimination" brought with them a number of additional questions, such as: whether the concept of equal protection rights should be translated into class or race rights; whether the rights created by the Fourteenth Amendment are personal rights guaranteed to the individual or should be turned into instruments of racial and class politics, thus introducing the concept of "racial-class" into the Constitution; whether the Court should extend its remedies against individual injury into remedies for discrimination identified solely by class; and whether affirmative action programs and quota systems call for the introduction of a racial code and special procedures for the racial classification of individuals. Justice John Paul Stevens warned in a dissenting opinion: "If the national government is to make a serious effort to define racial classes by criteria that can be administered objectively, it must study such precedents as the First Regulation of the Reich's Citizenship Law of 1935, which went into considerable detail in determining who met its standards of being a

'Jew.'" (*Fullilove v. Klutznick*, 448 U.S. 448, 534, 100 S.Ct. 2758, 2802 [1980].)

Decisions and legislative acts found inconsistent with new conditions or with new "public policy" are destined to be changed. In the endless process of testing and retesting ideas that might contribute to the well-being of our society, we will find that only the idea of natural law revealed to us through reason offers a guarantee to freedom and traditional morality. Laws and court decisions are just only if they conform to the standards of human rights and justice that exist independently of the legislative laws and court decisions based on them.

2. Common Law

As we mentioned earlier, Jefferson was greatly influenced by Locke's teaching on natural law. Its rules and maxims concerning immutable truth and justice also greatly influenced the adaptation of England's Common Law into American jurisprudence. In 1775 Burke asserted that nearly as many copies of Blackstone's *Commentaries* had been sold in the American colonies alone as in all England. The *Commentaries* gained such great popularity among the American colonists because of their lucidity and their orderly arrangement. In evaluating the impact of English Common Law on American law, however, it should be remembered that among some of the early settlers in this country there existed strong hostility toward the English Common Law.

This sentiment was intensely bitter among the

more radical Jeffersonian Democrats. They pointed
out that Blackstone, a royalist indebted to the government for his preeminence, glorified the principles of
that government. Such statements as "The King is not
only incapable of doing wrong but even of thinking
wrong; in him is no folly or weakness," as well as his
contempt for the common people and for the American colonists, did not earn him admiration among the
early settlers. They found the principles of England's
political mechanism, which restricted the participation of the people in general in government, inadmis-
sible in a democratic republic.

Jesse Root (1736-1822), Chief Justice of the Superior Court of Connecticut, also believed English
law, especially the Common Law, to be utterly inap-
plicable to American society. In his "The Origin of
Government and Laws in Connecticut" (1798), he argued that the settlers who emigrated from England to
America became free from any obligation of
"subjection" to the laws and jurisprudence of
England. Having stressed the difference in spirit and
principle between the laws of England and the laws of
Connecticut, he accused those who supposed that the
rules of the Common Law of England were the
common law of Connecticut of ignorance. The
"common law" of Connecticut, in his opinion, was derived from the law of nature while the English law
was derived from the vices of feudalism. Pointing to
the "rising empire of America," he wrote: "Our government and our rulers are from among ourselves;
chosen by the free, uninfluenced suffrages of
enlightened freemen; not to oppress and devour, but
to protect, feed, and bless the people, with the be-

nign and energetic influence of their power (as
ministers of God for good to them)." Consequently,
"these rights and liberties are our own, not holden by
the gift of a despot."

Referring to the system of government and jurisprudence based on the common law of Connecticut,
Jesse Root asked: what is common law? He found
three sources of this law. The first is what we would
call natural law. This aspect of common law is the
perfection of reason, "arising from the nature of God,
of man, and of things, and from their relations, de-
pendencies, and convictions." It is universal and ex-
tends to all men; it is superior to all other laws in that
all positive laws are to be construed by it, and
"wherein they are opposed to it they are void." Because of the dignity of its origin, and the sublimity of
its principles, "it is visible in the volume of nature, in
all the works and ways of God."

Secondly, Root saw in certain usages and customs
another brand of common law. The courts of justice,
he argued, should recognize rules derived from unwritten customs and regulations which have the sanction of universal consent and adoption and are in
practice among the citizens at large or among particular classes of men (such as farmers, merchants, etc.)
and declare them "to be obligatory upon the citizens
as necessary rules of construction and of justice."

The third source of common law, according to Root,
is the "adjudication of the courts of justice and the
rules of practice adopted by them." He bitterly
resented the idea of basing American jurisprudence
on foreign constitutions and laws. "Let us," wrote
Root, "duly appreciate our own government, laws and

manners and be what we profess—an independent nation—not to plume ourselves upon being humble imi-
tators of foreigners, at home and in our own country.
But let our manners in all respects be characteristic of
the spirit and principles of our independence."

Root was not alone in his resentment of the idea of
"borrowing" our legal system from the English.
Other legal scholars exalted the American genius for
"simplicity" over the "brutal, ferocious, and inhu-
man" British laws that exhibited "feudal" complexity.
Charles Jared Ingersoll (1782-1862), a prominent Philadelphia lawyer, called for complete independence from English example and believed that the
"use and respect of American jurisprudence in Great
Britain will begin only when we cease to prefer their
adjudication to our own." He claimed that the "harsh
doctrines of the Common Law have all been melted
down by the genial mildness of American institutions." ("A Discourse concerning the Influence of
America on the Mind," 1823.)

William Sampson (1764-1836), an Irish lawyer ex-
iled by the British government but welcomed by the
New York bar, was a confirmed and ferocious enemy
of English Common Law. In his "An Anniversary Discourse, 1824," he assigned to law a place next to religion because "it is a guide of all our actions, and the
rule of our conduct," and accused his generation of
having set aside the great example of self government
by continuing to serve a "pagan idol," called by the
"mystical and cabalistic name of Common Law."
Sampson saw no virtues in Common Law, and de-
rided what he saw as contradictory (but all overly
respectful) attitudes toward it—of those who main-

tained that it was perfect in its inception but became corrupted over time, as well as of those who believed that it had a barbarous origin but gradually grew to perfection. "It was common sense," Sampson wrote, in dispraise of those who praised the Common Law irrationally, "but of an artificial kind, such as is not the sense of any common man; it was the perfection of reason, but that meant artificial reason."

Sampson saw no reason why English judges should continue to legislate for the United States. They were appointed by a king who is "the fountain of their justice," and he argued that judicial decisions should not be imported from a despotic country, where the will of the sovereign is the law, to a free country, "whose history is yet unstained with crime and usurpation," where the "faithful chronicler of its short but bright career may invoke both liberty and truth to bear him company." Sampson is one of the most eloquent and spirited defenders of the idea of a native tradition of American law, and his essay repays study.

The flame of opposition against English Common Law, which started in Virginia in 1799 or 1800 in consequence of opposition to the Alien and Sedition Acts, spread to other states. Some of them were ready to abolish Common Law altogether; others, like Pennsylvania, prohibited the reading or quoting in courts of justice of postrevolutionary British authorities; still others saw in Common Law a code of a monarchical country that represented a danger to the institutions of the newly born free nation.

In spite of pleas for a fresh codification of American law, the leading legal scholars succeeded in subduing the native suspicion of Common Law. They were not impassioned nationalists, and they fully realized that some of our legal ideas were originally derived from the laws of England which, when adapted to our own situation and circumstances, had been incorporated into the American system. What they were seeking was an alliance between the restless spirit of freedom, which made man a free agent responsible in his opinions and actions only to his God, and the laws to which man is amenable. His rights, they believed, were not a gift of an arbitrary ruler but derived from his covenant with God. They were building a society whose laws proclaimed that nothing is so high as to be above their reach or so low as to be beyond their care. A new and original legal system combining Common Law with Civil was thus born to preserve the purity and protect the perpetuity of individual freedom. The new legal system demanded that every citizen should hold his life, liberty, property, and immunities under the protection of the general rules which govern a free society.

Joseph Story (1779-1845), with his *Commentaries*, and the famous legal analyst James Kent (1763-1847), with his *Commentaries on American Law*, were both powerful figures, recognized as the American equivalents to Blackstone, who defended the Common Law. It seemed to them that it was impossible to abolish the Common Law since it represented an accumulation of reasonable decisions and served (at that time) as a basis for jurisprudence in twenty-three of the twenty-four states of the Union (the exception was Louisiana). There was no need, in their opinion, to rush into a new codification or sudden innovations, when the administration of justice in the United

States could incorporate the laws of England by adapting them to the spirit of a free, independent nation.

Justice Story stressed as the "proudest attribute of American jurisprudence" the right of judicial tribunals to decide questions of constitutional law. While in other governments these questions could not be entertained by courts of justice, since the legislative authority was practically omnipotent, in the United States they became the subject of judicial inquiry. The American judiciary, he maintained, possesses no control over the purse or arms of the government; it can neither enact laws nor levy taxes and raise armies. But when our judges are "fearless and firm in discharge of their functions," stated Story, "popular leaders cannot possess a wide range of oppression, but must stand rebuked in their ambitious career for power."

According to Story, the role of our courts and of the rule of law is to prevent the deprivation of any legal right in violation of the fundamental constitutional guarantees and to serve as a bulwark against all arbitrary acts of power. One assertion by Story was a prophecy which we have seen fulfilled in totalitarian countries, to wit, whenever the liberties of a country are to be destroyed, "the first step in the conspiracy will be to bring courts of justice into odium; and, by overawing the timid, and removing the incorruptible, to break down the last barrier between the people and universal anarchy and despotism."

In Soviet Russia and in other totalitarian countries, the courts are the instruments of the party or the dictator in power, and the judicial process is subordinated to the political end of assuring the continued authority of the party or dictator. Thus the judicial process is used to legalize the use of terror by the State and its secret police and to sanction political terror within the context of formal legalism.

Chancellor Kent also believed that our freedom and our private rights depend upon the enlightened and faithful administration of justice and upon government through the rule of law. He saw in judicial power the "ultimate expounder of the Constitution." The rule of law, according to Kent, is not limited to mere knowledge of the laws since "knowledge alone is not sufficient for pure and lasting fame. It is mischievous, and even dangerous, unless it be regulated by moral principles." The rule of law demands sacred reverence for the truth and must be founded on principles of morality and justice.

The two great luminaries of nineteenth-century legal theory, Story and Kent, also believed that Christianity is a part of Common Law. There has never been a period, wrote Story, "in which the Common Law did not recognize Christianity as lying at its foundations." Christianity is a part of Common Law, "from which it seeks the sanction of its rights, and by which it endeavors to regulate its doctrines." The activities of the organized church which have at times brought disgrace upon it Story attributed to the "error" of the Common Law that tolerated nothing but Christianity as taught by its own established church, either Protestant or Catholic. By consigning with unrelenting severity the conscientious heretic to the stake, justice was debased, "and religion itself made the minister of crimes, by calling in the aid of secular

power to enforce the conformity of belief, whose rewards and punishment belong exclusively to God."

Kent expected a lawyer not only to be taught every branch of our own jurisprudence but also to be properly instructed "in moral science" and "to have his passions controlled by the discipline of Christian truth." Nations are bound by the same obligations of truth, justice, and humanity that bind individuals in private life. The law of nations, stated Kent, may not offer efficient sanctions, but its principles are "founded in the maxims of eternal truth, in the immutable law of moral obligations, and in the suggestions of an enlightened public interest."

The compelling forces of natural justice and of the moral and ethical principles which move forward under the dynamics of human history will bring changes that must be met not with resistance but with the desire to preserve continuity with the past. Disraeli once warned, "We must choose to be managers of change, or we will be victims of change."

The three sources of freedom we have been discussing—religion, education, and law—permit us to be managers of technological changes. They flourish best when they are in alliance with each other, adorn each other, and jointly offer support and protection to all human rights—and most of all to the right that the Creator has endowed on all human beings to be free and to be protected against tyranny.